T0275663

The Analytical Chemistry of Cannabis

Emerging Issues in Analytical Chemistry

Series Editor
Brian F. Thomas

AMSTERDAM • BOSTON • HEIDELBERG • LONDON
NEW YORK • OXFORD • PARIS • SAN DIEGO
SAN FRANCISCO • SINGAPORE • SYDNEY • TOKYO

The Analytical Chemistry of Cannabis

Quality Assessment, Assurance, and Regulation of Medicinal Marijuana and Cannabinoid Preparations

Brian F. Thomas
Analytical Chemistry and Pharmaceutics, RTI International, Research Triangle Park, NC, United States

Mahmoud A. ElSohly
National Center for Natural Products Research, Research Institute of Pharmaceutical Sciences and Department of Pharmaceutics, School of Pharmacy, University of Mississippi, MS, United States

AMSTERDAM • BOSTON • HEIDELBERG • LONDON
NEW YORK • OXFORD • PARIS • SAN DIEGO
SAN FRANCISCO • SINGAPORE • SYDNEY • TOKYO

Elsevier
Radarweg 29, PO Box 211, 1000 AE Amsterdam, Netherlands
The Boulevard, Langford Lane, Kidlington, Oxford OX5 1GB, UK
225 Wyman Street, Waltham, MA 02451, USA

Published in cooperation with RTI Press at RTI International, an independent, nonprofit research
institute that provides research, development, and technical services to government and
commercial clients worldwide (www.rti.org). RTI Press is RTI's open-access, peer-reviewed
publishing channel. RTI International is a trade name of Research Triangle Institute.

Notices
Knowledge and best practice in this field are constantly changing. As new research and
experience broaden our understanding, changes in research methods or professional practices,
may become necessary.

Practitioners and researchers must always rely on their own experience and knowledge in
evaluating and using any information or methods described herein. In using such information or
methods they should be mindful of their own safety and the safety of others, including parties for
whom they have a professional responsibility.

To the fullest extent of the law, neither the Publisher nor the authors, contributors, or editors,
assume any liability for any injury and/or damage to persons or property as a matter of products
liability, negligence or otherwise, or from any use or operation of any methods, products,
instructions, or ideas contained in the material herein.

ISBN: 978-0-12-804646-3

British Library Cataloguing-in-Publication Data
A catalogue record for this book is available from the British Library

Library of Congress Cataloging-in-Publication Data
A catalog record for this book is available from the Library of Congress

For Information on all Elsevier Publishing publications
visit our website at http://store.elsevier.com/

This book has been manufactured using Print On Demand technology.

DEDICATION

This work is dedicated to my wife Cathy, and my mentors Billy Martin, Ed Cook, Bob Jeffcoat, and Ken Davis.

CONTENTS

LIST OF CONTRIBUTORS

Suman Chandra
National Center for Natural Products Research, Research Institute of Pharmaceutical Sciences, School of Pharmacy, University of Mississippi, MS, United States

Mahmoud A. ElSohly
National Center for Natural Products Research, Research Institute of Pharmaceutical Sciences and Department of Pharmaceutics, School of Pharmacy, University of Mississippi, MS, United States

Michelle Glass
Department of Pharmacology, University of Auckland, Auckland, New Zealand

Hemant Lata
National Center for Natural Products Research, Research Institute of Pharmaceutical Sciences, School of Pharmacy, University of Mississippi, MS, United States

Raphael Mechoulam
Institute for Drug Research, The Hebrew University of Jerusalem, Jerusalem, Israel

Roger G. Pertwee
Institute of Medical Sciences, University of Aberdeen, Aberdeen, Scotland, United Kingdom

Brian F. Thomas
Analytical Chemistry and Pharmaceutics, RTI International, Research Triangle Park, NC, United States

Ryan G. Vandrey
Behavioral Biology Research Unit, Johns Hopkins University School of Medicine, Baltimore, MD, United States

FOREWORD

Cannabis has been used for thousands of years for recreational, medicinal, or religious purposes. However, the determination of the chemical structures of its cannabinoids, terpenes, and many other constituents, and of the pharmacological actions and possible therapeutic uses of some of these compounds, began less than 100 years ago. This book begins by describing the cultivation, harvesting, and botanical classification of cannabis plants, and then goes on to specify how these plants produce some of their chemical constituents. Subsequent chapters focus on medical formulations of cannabis and cannabis-derived drugs, on the routes of administration of these formulations, and on analytical methods that are used in the formulation development and for the quality control or stability assessment of cannabis constituents. The penultimate chapter deals with regulatory and additional formulation-related issues for medical cannabis and cannabinoids, while the final chapter identifies ways in which analytical chemistry will most likely contribute to the development of cannabinoid therapeutics in the future. This book provides much needed insights into the important roles that analytical chemistry has already played and is likely to continue to play in the development of cannabis and its constituents as medicines.

Roger G. Pertwee MA, DPhil, DSc, HonFBPhS
Institute of Medical Sciences
University of Aberdeen
Aberdeen, Scotland, United Kingdom

Pharmacology began with natural products and, over some years on either side of 1900, evolved into a rigorous scientific discipline dominated, at least in the West, by well-defined chemical entities, either extracted and processed or synthesized. The two traditions evolved together, each informing the other, the natural strain by long experience pointing the way toward how a drug development program might be structured, the synthetic strain contributing molecular specificity, with analytical chemistry a common element. The resultant contribution to modern medicine, with all its caveats and controversies, must be accounted as one of the great advancements in science.

Natural products pharmacology is very much alive. However, that "natural" is one cause of the popular misconception that herbs are in some way better or safer than pills. Though some herbal remedies do appear to be safe and effective, the opposite is closer to the truth. Cannabis is a good example. The number of parameters on which cannabis products can vary is enormous, from strain, growing conditions, harvesting methods, and handling to storage and processing of the raw material to combination with a wide variety of foods and other excipients in manufacturing to methods of administration (eating, smoking, "vaping," applying to mucous membranes). At every step, from planting through consumption, myriad influences can alter dose, absorption rate, interactions among constituents, exposure to toxins, and a host of other factors that can result in underdosing, overdosing, and various types and levels of acute and chronic poisoning, not excepting an increase in the probability of lung cancer. Even if quality were well controlled, which on the whole is very much not the case, this complexity means that governmental oversight of cannabis products cannot be as close and complete as that for prescription and over-the-counter pharmaceuticals. Caveat emptor.

Governments around the world are coming slowly to the conclusion that, in the absence of draconian enforcement, and to a nontrivial extent in its presence, people are going to use cannabis for medicine and recreation. The Internet spreads knowledge of genetic sequencing, metabolomics, proteomics, and other disciplines such that people are going to manipulate cannabis, as they have long done by selective breeding, to maximize its mental and physical effects and tailor the quality of those effects. The present legal status in the United States and elsewhere cannot stop these activities by amateurs, but it does inhibit research by professionals to investigate the basic science of cannabis, and to use this information to better understand neurophysiological function, develop new medicines for people and animals, and find ways to deal with cannabis addiction. Tight control of marijuana and inhibition of legal research has arguably led to another paradoxical effect: driving the chemistry underground, which has resulted in the proliferation of new and more dangerous synthetic cannabinoids. There needs to be more involvement by elements of the US Food and Drug Administration rather than the Drug Enforcement Agency.

Clearly, the policy, regulatory, and research challenges that accompany the study and understanding of cannabis are unique. Despite all the issues, research continues, understanding of cannabis and its effects is evolving, policies are in flux, and the literature is ever-changing. The aim of this book is to provide the reader with a detailed understanding of the analytical chemistry of cannabis and cannabinoids as the foundation for quality, safety, and utility of cannabis-derived therapeutics, and offer direction for future advancements.

ACKNOWLEDGMENTS

The authors thank RTI Press and RTI International for their support of this project, as well as the continued support of RTI research on cannabis over the years by the National Institute on Drug Abuse.

We appreciate the opportunity to work with the editorial and production team at Elsevier—Katy Morrissey, Amy Clark, Vijayaraj Purushothaman, and the many who go unmentioned—in bringing this first volume in the series "Emerging Issues in Analytical Chemistry" to fruition.

Chapter 1, "The Botany of *Cannabis sativa* L.," was prepared collectively by Dr Suman Chandra, Dr Hemant Lata, and Dr Mahmoud A. ElSohly at the University of Mississippi, whose work was supported in part with federal funds from the National Institute on Drug Abuse, National Institutes of Health, Department of Health and Human Services, USA, under contract No. N01DA-10-7773.

Thanks to Dayle G. Johnson of RTI International for the cover design.

We are especially indebted to Dr Gerald T. Pollard for his editorial assistance. His attention to detail and overall project management are greatly appreciated.

The Botany of *Cannabis sativa* L.

Cannabis sativa L. is a widespread species in nature. It is found in various habitats ranging from sea level to the temperate and alpine foothills of the Himalayas, from where it was probably spread over the last 10,000 years.[1,2] The age-old cultivation makes its original distribution difficult to pinpoint.[3] Cannabis has a long history of medicinal use in the Middle East and Asia, with references as far back as the 6th century BCE, and it was introduced in Western Europe as a medicine in the early 19th century to treat epilepsy, tetanus, rheumatism, migraine, asthma, trigeminal neuralgia, fatigue, and insomnia.[4,5]

As a plant, it is valued for its hallucinogenic and medicinal properties, more recently being used for pain, glaucoma, nausea, asthma, depression, insomnia, and neuralgia.[6,7] Derivatives are used in HIV/AIDS[8] and multiple sclerosis.[9] The pharmacology and therapeutic efficacy of cannabis preparations and its main active constituent Δ^9-tetrahydrocannabinol (Δ^9-THC) have been extensively reviewed.[10–12] The other important cannabinoid constituent of current interest is cannabidiol (CBD). There has been a significant interest in CBD over the last few years because of its reported activity as an antiepileptic agent, particularly its promise for the treatment of intractable pediatric epilepsy.[13,14] Other than Δ^9-THC and CBD, tetrahydrocannabivarin (THCV), cannabinol (CBN), cannabigerol (CBG), and cannabichromene (CBC) are major isolates. Fig. 1.1 shows chemical structures.

Cannabis is also one of the oldest sources of food and textile fiber.[15–17] Hemp grown for fiber was introduced in Western Asia and Egypt and subsequently in Europe between 1000 and 2000 BCE. Cultivation of hemp in Europe became widespread after 500 CE. The crop was first brought to South America (Chile) in 1545, and to North America (Port Royal, Acadia) in 1606.[18] Now its cultivation is prohibited or highly regulated in the United States.

The Analytical Chemistry of Cannabis. DOI: http://dx.doi.org/10.1016/B978-0-12-804646-3.00001-1

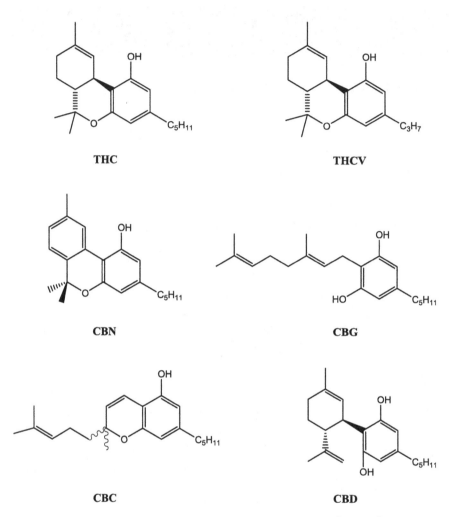

Figure 1.1 Chemical structures of major cannabinoids present in Cannabis sativa. *Δ^9-THC, Δ^9-tetrahydrocan-nabinol;* THCV, *tetrahydrocannabivarin;* CBN, *cannabinol;* CBG, *cannabigerol;* CBC, *cannabichromene;* CBD, *cannabidiol.*

BOTANICAL DESCRIPTION

Table 1.1 describes the botanical nomenclature of *C. sativa* L. Cannabis is a highly variable species in terms of botany, genetics, and chemical constituents. The number of species in the *Cannabis* genus has long been controversial. Some reports proposed *Cannabis* as a polytypic genus.[19–22] However, based on morphological, anatomical, phyto-chemical, and genetic studies, it is generally treated as having a single, highly polymorphic species, *C. sativa* L.[23–26] Other reported species

Table 1.1 Botanical Nomenclature of *Cannabis sativa* L.	
Category	Botanical Nomenclature
Kingdom	Plantae—Plants
Subkingdom	Tracheobionta—Vascular plants
Superdivision	Spermatophyta—Seed plants
Division	Magnoliophyta—Flowering plants
Class	Magnoliopsida—Dicotyledons
Subclass	Hamamelididae
Order	Urticales
Family	Cannabaceae
Genus	*Cannabis*
Species	*Cannabis sativa* L.

are *Cannabis indica* Lam. and *Cannabis ruderalis* Janisch, but plants considered to have belonged to these species are now recognized as varieties of *C. sativa* L. (var. *indica* and var. *ruderalis*, respectively). *Cannabis sativa* and *indica* are widely cultivated and economically important; *Cannabis ruderalis* is hardier and grows in the northern Himalayas and the southern states of the former Soviet Union but is rarely cultivated for drug content.

The main morphological difference between *indica* and *sativa* is in their leaves. The leaves of *sativa* are much smaller and thinner, whereas those of *indica* have wide fingers and are deep green, often tinged with purple; at maturity, they turn dark purple. *Indica* plants are shorter and bushier, usually under 6 ft tall and rarely over 8 ft *Indica* has short branches laden with thick, dense buds, which mature early, usually at the beginning of September in the Northern Hemisphere. *Indica* buds also vary in color from dark green to purple, with cooler conditions inducing more intense coloration. *Indica* flowers earlier. The natural distribution of *indica* is Afghanistan, Pakistan, India, and surrounding areas. The plants of *sativa* have long branches, with the lower ones spreading 4 ft or more from the central stalk, as on a conical Christmas tree. Height varies from 6 ft to more than 20 ft, with 8−12 ft being the most common. Buds are long and thin and far less densely populated than in *indica*, but longer, sometimes 3 ft or more. Maturation time varies considerably depending on the variety and environmental conditions. Low Δ^9-THC Midwestern *sativa* varieties (ditchweed) mature in August and September, while equatorial varieties mature from October to December. Buds of *sativa* require intense light to thicken and swell; *indica* does not. *Sativa* tends to be higher in Δ^9-THC and lower in CBD than *indica*. S*ativa* is found all over the world and comprises most of the drug type equatorial varieties such as Colombian, Mexican, Nigerian, and South African, where marijuana plants can be very potent. Cannabis has many local common names, some of which are given in Table 1.2.

Normally, cannabis exhibits a dioecious (male and female flowers develop on separate plants) and occasionally a monoecious (hermaphrodite) phenotype. It flowers in the shorter days (below 12-h photoperiod) and continues growing vegetatively in the longer photoperiod. Sex is determined by heteromorphic chromosomes (males being heterogametic XY, females homogametic XX). Male flowers can be differentiated from female by their different morphological appearance. At the vegetative stage, differentiation is difficult because of morphological similarities. Molecular techniques, however, can differentiate at an early stage.[27–32]

Cannabis is wind pollinated. For the production of cannabinoids (or phytocannabinoids), female plants are preferred for several reasons. First, they produce higher amounts of cannabinoids. Second, once pollinated, female plants produce seeds at maturity, whereas seed-free

Table 1.2 Common Cannabis Names in Different Languages

Language	Common Names
Arabic	Bhang, hashish qinnib, hasheesh kenneb, qinnib, tîl
Chinese	Xian ma, ye ma
Danish	Hemp
Dutch	Hennep
English	Hemp, marihuana
Finnish	Hamppu
French	Chanvre, chanvre d'Inde, chanvre indien, chanvrier
German	Hanf, haschisch, indischer hanf
Hindi	Bhang, charas, ganja
Japanese	Mashinin
Nepalese	Charas, gajiimaa, gaanjaa
Portuguese	Cânhamo, maconha
Russian	Kannabis sativa
Spanish	Cáñamo, grifa, hachís, mariguana, marijuana
Swedish	Porkanchaa

plants (sinsemilla, a Spanish word) are preferred for their higher yield of secondary metabolites. Third, if several cannabis varieties are being grown together, cross-pollination would affect the quality (chemical profile) of the final product. To avoid this, removing male plants as they appear, screening female clones for higher metabolite content, and conservation and multiplication using biotechnological tools ensures the consistency in chemical profile that is desirable for pharmaceuticals.

CHEMICAL CONSTITUENTS AND PHENOTYPES OF *C. SATIVA* L.

CBN was the first cannabinoid to be isolated[33,34] and identified[35–37] from *C. sativa*. The elucidation of CBN led to speculation that the psychotropically active constituents of cannabis could be THCs. The nonpsychotropic compound CBD was subsequently isolated from Mexican marijuana[38] and the structure was determined.[39] Gaoni and Mechoulam, two pioneers of cannabis research, determined the structure of Δ^9-THC after finally succeeding in isolating and purifying this elusive compound (see Mechoulam Close-up: How to Pamper an Idea).[40] Since then, the number of cannabinoids and other compounds isolated from cannabis has increased continually, with 545 now reported. Of these, 104 are phytocannabinoids (Table 1.3). From the isolation and structural elucidation of Δ^9-THC in 1964 until 1980,

No.	Groups	Number of Known Compounds
\multicolumn{3}{l}{**Table 1.3 Constituents of *Cannabis sativa* L.**}
1	CBG type	17
2	CBC type	8
3	CBD type	8
4	Δ^9-THC type	18
5	Δ^8-THC type	2
6	CBL type	3
7	CBE type	5
8	CBN type	10
9	CBND type	2
10	CBT type	9
11	Miscellaneous	22
12	Total cannabinoids	104
13	Total noncannabinoids	441
	Total	545

61 phytocannabinoids were isolated and reported.[41] Only nine new ones were characterized between 1981 and 2005,[42] but 31 were reported between 2006 and 2010. The 13 chemical constituent type groups shown in Table 1.3 suggests the chemical complexity of the cannabis plant.[42]

The concentration of Δ^9-THC and CBD, the most abundant cannabinoids, can be characterized qualitatively and quantitatively.[43] Qualitative characterization is based on the Δ^9-THC/CBD ratio and assigning the plant to a discrete chemical phenotype (chemotype). In 1971, cannabis was initially characterized in two phenotypes, drug type and fiber type, by Fetterman et al.[44] A Δ^9-THC/CBD ratio >1 was drug type, a lesser ratio was fiber type. In 1973, Small and Beckstead proposed three categories based on the ratio: drug type if >1, intermediate if close to 1, and fiber if <1.[45,46] In 1987, Fournier et al. added a rare chemotype that was characterized by a very low content of Δ^9-THC and CBD with CBG as the predominant constituent.[47]

Quantitatively, the plant is characterized by potency through measuring the level of its most abundant cannabinoids, Δ^9-THC and CBD, in its tissues (Fig. 1.2). The levels of cannabinoids are controlled by the interaction of several genes and also influenced by the growth environment of the plant.[48−52] Numerous biotic and abiotic factors affect cannabinoid production, including the sex, growth stage, environmental parameters, and fertilization.[23,50,53−56]

CANNABIS BIOSYNTHESIS

Fig. 1.3 is a schematic of cannabinoid biosynthesis. In the plant, Δ^9-THC, CBD, and CBC are in their acid forms.[57−59] Two independent pathways, cytosolic mevalonate and plastidial methylerythritol phosphate (MEP), are responsible for terpenoid biosynthesis. The MEP pathway is reported to be responsible for the biosynthesis of the terpenoid moiety.[12] Olivetolic acid (OLA) and geranyl diphosphate (GPP) are derived from the polyketide and the deoxyxylulose phosphate (DOXP)/MEP pathways, respectively, followed by condensation under the influence of the prenylase olivetolate geranyltransferase, yielding cannabigerolic acid (CBGA). CBGA, in turn, is oxidocyclized

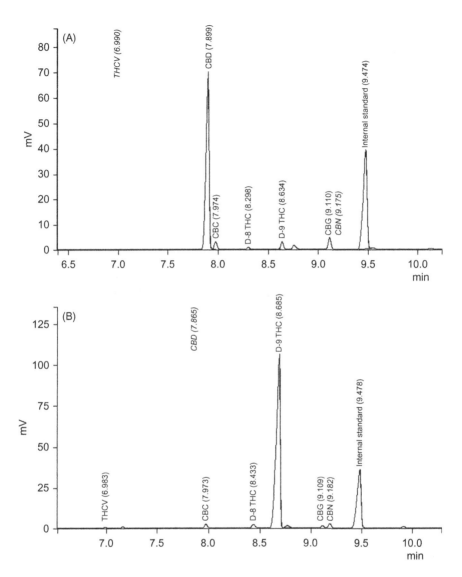

Figure 1.2 Gas chromatography-flame ionization detector (GC-FID) analysis of (A) a high CBD type and (B) a high Δ⁹-THC type cannabis plant.

by flavin adenine dinucleotide-dependent oxidases, namely, cannabi-chromenic acid (CBCA) synthase, cannabidiolic acid (CBDA) synthase, and Δ^9-THCA synthase, producing CBCA, CBDA, and Δ^9-THCA, respectively.[60,61]

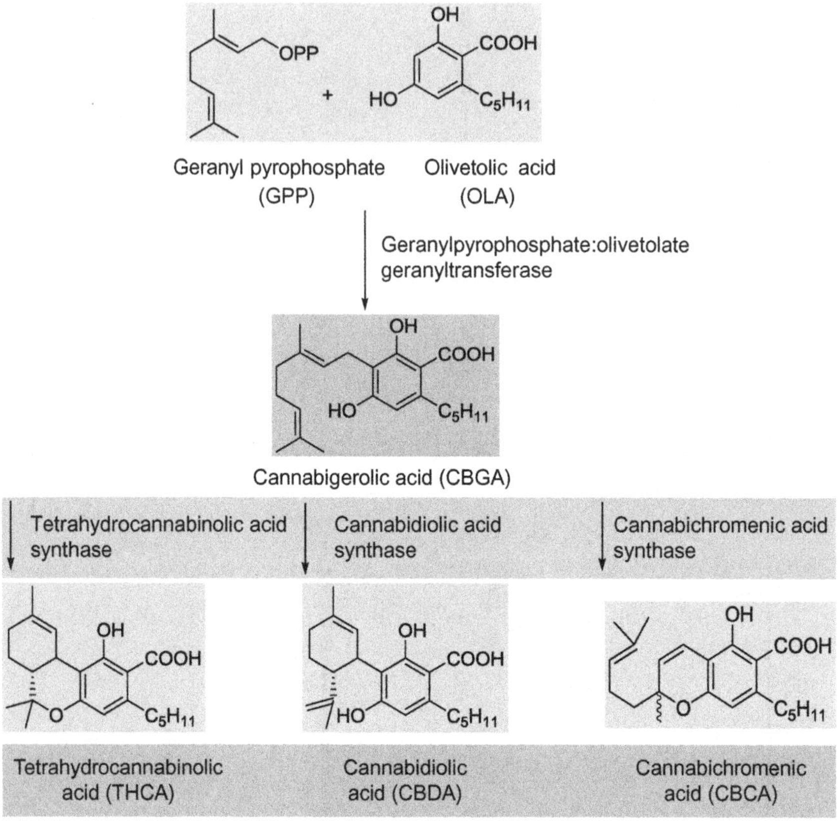

Figure 1.3 Biosynthetic pathway of tetrahydrocannabinolic acid, cannabidiolic acid, and cannabichromenic acid.

SELECTION OF ELITE CLONES FOR PLANT PROPAGATION

The quality, safety, and efficacy of starting material are basic prerequisites in the pharmaceutical industry. Cannabis as a feedstock is more challenging because it is a chemically complex and highly variable plant due to its allogamous nature. The chemical composition of cannabis biomass is affected by a range of factors such as genetics, environment, growth conditions, and harvesting stage. Therefore, selection of elite starting material (female clone) based on chemical composition, conservation, and mass multiplication using advanced biotechnological tools is a suitable way to ensure the consistency in chemical profile of a crop for pharmaceuticals.

In our laboratory, we developed a GC-FID method for screening and selection of elite biomass based on major cannabinoid content.

Briefly, quantitative analysis of seven major cannabinoids (THCV, CBD, CBC, Δ^8-THC, Δ^9-THC, CBG, and CBN) is done by solvent extraction followed by analysis using capillary gas chromatography, a method offering short analysis time and resolution of all cannabinoids on a single column. Three samples (100 mg each) are used for analysis from each manicured biomass sample. A 3-mL internal standard (IS) extraction solvent (100 mg of 4-androstene-3,17-dione + 10 mL chloroform + 90 mL methanol) is added to the sample and allowed to rest at room temperature for 1 h. The extract is then filtered through a cotton plug and the clear filtered material is transferred to an autosampler vial. Samples are placed onto the GC instrument along with vials of ethanol, internal standard/Δ^9-THC mixture (unextracted standard), and controls. The results are calculated by obtaining an average percentage of each cannabinoid from the two chromatograms of each sample. It must be noted that the response factor for the cannabinoids relative to IS is 1. Therefore, the area of each cannabinoid divided by that of the IS multiplied by the amount of IS added (3 mg) gives the percentage of each cannabinoid in the sample, as 100 mg of sample is used for analysis. For example, a cannabinoid with the same peak area as that of the IS represents a 3% concentration in the sample. The method has been validated to meet FDA-GMP requirements.

Once a female cannabis plant is screened and selected based on its cannabinoids profile, it can be used as a mother stalk for future propagation.

PLANT GROWTH AND CULTIVATION

Cannabis is an annual species and can be grown from seed or vegetative cuttings under indoor and outdoor conditions. Indoor cultivation under controlled environmental conditions can generate three or four crops per year, depending upon required per-plant biomass yield; outdoor cultivation is limited to one crop per year. Selection of starting material or variety depends upon the composition of active ingredients required in the end product.

Propagation Through Seed

Fig. 1.4 shows the typical seeds of a high-Δ^9-THC-yielding Mexican variety of *C. sativa* L. Cultivation through seed is an ancient and

Figure 1.4 Cannabis sativa *seeds (high-yielding Mexican variety).*

traditional method. Seeds can be sown in small biodegradable jiffy pots containing soil with good aeration and should be kept moist by watering with a light spray when the upper surface begins to feel dry. During winter, a heat mat can be used below the pots to increase the soil temperature and enhance germination.

Normally seeds start sprouting by the fourth day of plantation, and most of the viable seeds germinate within 15 days. Variation in the rate of seed germination depends on the variety, seed age, storage conditions, and soil and water temperatures. Germinated seedlings can be kept under cool fluorescent light with an 18-h photoperiod until the seedlings are big enough to transplant to bigger pots. Once transplanted, they can be kept under full-spectrum grow light (1000 W high-pressure sodium or metal-halide bulbs) with an 18-h photoperiod for further vegetative growth.

Upon completion of desired vegetative growth, plants may be exposed to a 12-h photoperiod for flowering. (At this stage, cuttings of selected healthy plants can be made and maintained at vegetative stage for screening purposes.) Onset of flowering normally occurs in 2 weeks, depending upon variety. Being a dioceous species, seed raised plants normally turn 50% male and 50% female, depending on the variety. Onset of male flowers normally occurs a week earlier than female flowers. For the production of useful secondary metabolites, female plants are preferred as they produce higher cannabinoid content. At this early flowering stage, male plants can be identified and separated from female plants. Once all the male plants are removed, female plants can be grown to full maturity for the production of sinsemilla (seedless) buds. Mature buds can be analyzed for cannabinoid content using GC-FID. Based on this analysis, elite high-yielding clones can be identified and their vegetative backup cuttings can be used as mother plants for future propagation.

Vegetative Propagation in Soil

The pharmaceutical industry requires consistency in the active ingredients of source material. Using cannabis as a source raw material remains especially challenging. In spite of being grown from seeds derived from a single cannabis mother plant, a considerable degree of variation in chemical composition of juvenile plants may be observed. Therefore, screening of high-yielding female plants and mass propagation of vegetative clones is the most suitable way to meet the demand for uniformity of the final product.

Once a best candidate female clone with a specific chemical profile is screened and selected, a fresh nodal segment about 6 to 10 cm long containing at least two nodes can be used for vegetative propagation. A soft apical branch is cut at a 45-degree angle just below a node and immediately dipped in distilled water. The base of the cutting is subsequently dipped in rooting hormone and planted in biodegradable jiffy pots (2 × 2 in) containing coco natural growth medium and a mixture (1:1) of sterile potting mix and fertilome. At least one node is covered by soil for efficient rooting. Plants are regularly irrigated and kept under controlled environmental conditions. Rooting initiates in 2 to 3 weeks, followed by transplantation to bigger pots after 5 to 6 weeks. The cuttings can be maintained at constant vegetative state under 18-h photoperiod (Fig. 1.5).

Vegetative Propagation in Hydroponics

Vegetative cuttings can also be grown in a hydroponics system. A small branch consisting of a growing tip with two or three leaves is cut and immediately dipped in distilled water. The base of the cutting is dipped in rooting hormone and inserted (~1 in) deep into a rockwool cube or a hydrotone clay ball supporting medium. Plants are supplied with vegetative fertilizer formula and exposed to fluorescent light under 18-h photoperiod. Rooting initiates in 2 to 3 weeks.

(A)

(B)

Figure 1.5 Indoor vegetative propagation of Cannabis sativa. *(A) Vegetative cuttings under fluorescent lights, (B) plant growing under full-spectrum metal-halide lamps.*

In Vitro Plant Regeneration

Tissue culture methods offer an alternative means of vegetative propagation. Clonal propagation through tissue culture, commonly called micropropagation, can be achieved in a short time and a small space. It is possible to produce plants in large numbers starting from a single clone. However, the process involves several stages, from initiation and establishment of aseptic cultures to multiplication, rooting of regenerated shoots, and hardening in soil. Direct organogenesis is the most reliable method for clonal propagation because it upholds genetic uniformity among progenies.[62-64] An efficient micropropagation protocol for mass growing of drug type varieties using apical nodal segments containing axillary buds[65-67] (Fig. 1.6), as well as the micropropagation of a hemp variety using shoot tips,[68] have been reported. Recently, our group developed an improved one-step micropropagation protocol using meta-topolin.[69]

Plant tissue culture is also considered the most efficient technology for crop improvement by the production of somaclonal

Figure 1.6 Micropropagation of Cannabis sativa. *(A) A representative mother plant, (B and C) fully rooted cannabis plants, (D) micropropagated plants under the acclimatization condition, and (E−G) well-established micropropagated plants in soil.*

and gametoclonal variants. The callus-mediated cultures have inheritable characteristics different from those of parent plants due to the possibility of somaclonal variability,[70] which may lead to the development of commercially important improved varieties. Micropropagation through callus production has been reported, including production of roots through cannabis calli,[71] occasional shoot regeneration,[63] and high-frequency plant regeneration from leaf tissue—derived calli.[72]

Quality Control of In Vitro Regenerated Plants

The sustainability of the regeneration systems depends upon the maintenance of the genetic integrity of micropropagated plants. Despite its potential, in vitro techniques are known to induce somaclonal variations. Further, the frequency of these variations varies with the source of explants and their regeneration pattern, media composition, and cultural conditions. Tissue culture—induced variations can be determined at the morphological, cytological, biochemical, and molecular levels with several techniques. At present, molecular markers are powerful tools used in the analysis of genetic fidelity of in vitro propagated plantlets. These are not influenced by environmental factors and generate reliable and reproducible results.

In our laboratory, DNA-based intersimple sequence repeat (ISSR) markers have been successfully used to monitor the genetic stability of the micropropagated plants of C. sativa.[73,74] Fully mature in vitro propagated plants were also analyzed for their chemical profile and cannabinoid content, and compared with mother plants and vegetatively grown plants from the same mother plant using GC-FID for quality assurance.[75]

Our results showed that micropropagated plants were highly comparable to the mother plant and vegetatively grown plants in terms of genetics, chemical profile, and cannabinoid content. These results confirm the clonal fidelity of in vitro propagated plants and suggest that the biochemical mechanism used to produce the micropropagated plants does not affect genetics or metabolic content. So these protocols can be used for mass propagation of true to type plants of C. sativa for commercial pharmaceutical use.

Conservation of Elite Germplasm

The conservation of plant genetic resources is vital for the maintenance and improvement of existing gene pool and plant breeding programs.[76] In the last few decades, in situ conservation methods have played an important role in the conservation of elite plant germplasm. We have developed protocols for the conservation of elite C. sativa clones using vegetative propagation, slow growth conservation techniques, and alginate encapsulation.[66,77]

INDOOR CULTIVATION

Indoor cultivation under controlled environmental conditions allows total control of the plant life cycle and the quality and quantity of the biomass as starting material for the production of a desirable cannabinoid profile for pharmaceutical use. Parameters such as light (intensity and photoperiod), temperature, carbon dioxide level, air circulation, irrigation, relative humidity, and plant nutrition are the most important factors.

Light

Light is a vital component for the photosynthesis in plants. Suitable light quality, optimum light intensity, and desirable photoperiod are important in the indoor cultivation of cannabis. Our study shows that cannabis plants can use high photosynthetic photon flux density ($\sim$1500 μmol/m^2/s) for efficient gas and water vapor exchange between leaves and the environment.[50,78] Different light sources can be used: fluorescent (for cuttings), metal-halide bulbs, high-pressure sodium lamps (for well-established plants), conventional bulbs, and light emitting diodes. However, with indoor lighting, it is difficult to match photosynthetically active radiation received in the bright outdoor sunlight. An 18-h photoperiod is optimum for vegetative growth; 12-h is recommended for the initiation of flowering.

Irrigation and Humidity

The amount of water and the frequency of watering vary with the growth stage, size of the plants and pots, growth temperature, humidity, and many other factors. During the early seedling or vegetative stage, keeping the soil moist is recommended. Once the plants are established, the top layer of soil must be allowed to dry out before the plants are watered again. Overall, the soil should not be kept constantly wet and the plant should not be allowed to wilt. In general, watering should be done based on the requirement of the plant depending on its growth stage and the size of the container. Vegetative cuttings require regular moisture on the leaves to maintain a high humidity in its microclimate until the plants are well rooted. Humidity around 75% is recommended during the juvenile stage and about 55% to 60% during the active vegetative and flowering stages.

Temperature

Depending upon the original growth habitat and the genetic makeup, the temperature response of photosynthesis varies with the cannabis variety. Growth temperature of 25°C to 30°C is ideal for most varieties.[50,51,79]

Level of Carbon Dioxide

Air circulation in the growing room is another important factor and is necessary for indoor cultivation of healthy plants. An elevated CO_2 level enhances photosynthetic carbon assimilation and so may accelerate growth and improve productivity. A doubling of CO_2 concentration has been reported to increase the crop yield by 30% or more in experiments conducted under close environmental conditions such as in greenhouses and growth chambers.[80-84] Doubling of CO_2 concentration ($\sim$750 μmol mol^{-1}) was reported to stimulate the rate of photosynthesis and water use efficiency by 50% and 111%, respectively, as compared to ambient CO_2 concentration.[50,85] Therefore, supplementing CO_2 and proper air circulation in the growing room are recommended and will scale up the rate of photosynthesis and overall growth.

Plant Nutrition

Cannabis requires a minimum amount of nitrogen and a high level of phosphorus to promote early root growth. For vegetative growth, a higher level of nitrogen is required. For flowering, more potassium and phosphorous for the production of buds is needed.

OUTDOOR CULTIVATION

Cannabis is an annual. In the natural environment, it flowers at the end of summer (shortening days) irrespective of planting date or plant age. Seed is set before the arrival of winter and plant dies, if not harvested. Outdoor planting normally starts during late March or early April, depending on weather conditions, and could last into November or early December for some varieties. Starting from seed, plants may be raised in small jiffy pots and the selected healthy seedlings transplanted to the field (seeds also may be directly planted in the field). Male flowers start appearing within 2 to 3 months, around the middle of July, followed by female flowers. Male plants are generally removed from the fields for reasons stated above. Vegetative propagation of selected elite female clones and their field plantation is generally preferred over seedlings for the consistency in the chemical profile of the end product. Similar to seedlings, propagation of cuttings can be done in small jiffy pots and rooted cuttings then planted in the field by hand or by automated planters. Fig. 1.7 shows a typical outdoor cultivation through vegetative cuttings.

Throughout the growing season, a few randomly selected plants from different plots are periodically analyzed for cannabinoid content. We found that the amount of Δ^9-THC increases with the age of the plant, reaching the highest level at the budding stage and plateauing before the onset of senescence. The maturity of the crop is determined

Figure 1.7 Outdoor cultivation of cannabis crop. (A and B) plants at vegetative stage, (C and D) plants at budding stage.

visually and confirmed using GC-FID based on the Δ^9-THC and other cannabinoid content in samples collected at different growth stages. Since the whole plant does not mature at the same time, mature upper buds are harvested first and other branches are given more time to achieve maturity.

Outdoor cultivation has a few advantages and disadvantages in comparison to indoor. Field-cultivated plants are normally bigger and have higher biomass. Growing in the natural environment does not require the intensive investment in equipment and maintenance that indoor does. The primary disadvantage is less control over growing conditions. The weather may be unfavorable for harvest when the plants have reached maturity, or a thunderstorm may seriously damage plants when they are ready to harvest. Outdoor plants need a longer growing season than indoor.

HARVESTING AND PROCESSING

Fig. 1.8 shows harvesting, drying, and processing of field-grown cannabis biomass.

Figure 1.8 Harvesting, drying, and processing cannabis biomass. (A) Harvesting mature plants, (B) drying biomass, (C) dried cannabis buds, and (D) processed plant material in barrel.

Harvesting

Determining the optimum harvesting stage is a critical step in cannabis cultivation. Too early or too late can significantly affect the yield of Δ^9-THC. Periodical monitoring of Δ^9-THC level allows harvesting material with the desired content. Harvesting should be done in the morning because Δ^9-THC level peaks before noon and then gradually declines. Within the plant, the mature top buds may be harvested first and the others allowed time to mature.

Handling, Drying, Processing, and Storage

During harvest, drying, and processing, gloves are recommended. If the biomass is being used as starting material for pharmaceuticals, contact with the ground should be avoided. Dry and large leaves may be removed from mature buds before drying (Fig. 1.8).

The drying facility is based on the size of cultivation. For large-scale growing, a commercial tobacco drying barn (such as BulkTobac, Gas-Fired Products, Inc., Charlotte, NC) can be used. For small samples, a simple laboratory oven will suffice for overnight drying at 40°C.

When the material is dried, it can be hand manicured. Big leaves should be separated from buds. The buds can be gently rubbed through screens of different sizes to separate small stems and seeds (if any) from the dried biomass. Automated machines designed for plant processing can separate big stems and seeds from the useable biomass.

Storing Biomass

Properly dried and processed biomass can be stored in FDA-approved sealed fiber drums containing polyethylene bags at 18°C to 20°C for the short term. For the long term, -10°C in a freezer is recommended. Stability of Δ^9-THC and other cannabinoids in biomass and products has been reviewed by several authors.[86-88]

CONCLUSION

Cannabis is very special in the plant kingdom in that it belongs to a family (Cannabaceae) with a single genus (*Cannabis*) with only one species (*sativa*) that has many varieties. The plant is very rich in constituents, the most specific of which are the cannabinoids that have not been reported in any other plant, and it has broad pharmacological properties with tremendous medical potential in the treatment of epilepsy, spasticity, inflammation, irritable bowel syndrome, pain, and other disorders. Methods for growing, harvesting, processing, formulation, and use continue to evolve towards an important position in the pharmacopeia.

Close-Up: How to Pamper an Idea

R. Mechoulam
Hebrew University, Institute for Drug Research, Jerusalem, Israel

A scientific idea has not only to be pursued, but also pampered—in particular, if it is not welcome.

In the early 1960s as a newly appointed junior scientist at the Weizmann Institute in Rehovot, I was supposed to work with the head of my department, but could also explore some ideas of my own. I was interested in the chemistry and biological effects of natural products. I looked for biologically active plants that had not been previously well investigated. To my great surprise, I noted that the chemistry and hence the various activities of *Cannabis sativa*, the hashish plant, were not well known.

Morphine and cocaine had been isolated in the 19th century, and the availability of these alkaloids had made possible biochemical, pharmacological and clinical work with them. Why not cannabis? Actually its chemistry and biology had been investigated. There were numerous publications in the 19th century in mostly obscure journals. More recently, in the late 1930s and early 1940s, Roger Adams, a prominent US chemist, and Lord Alexander Todd, a Nobel Prize winner, had looked at the chemistry of cannabis, but apparently the active constituents had never been isolated in pure form and no definite structures had been put forward. The reasons may have been technical. We know today that the cannabinoids—a term I coined some years later—are present in the plant as a mixture of constituents with closely related chemical structures which presumably could not be separated by the methods available then.

Later, legal obstacles made work on cannabis almost impossible, particularly in North America and Europe. Cannabis—an illicit entity—was not readily available to most scientists, and research with it was next to impossible for academics, as few of them could follow effectively the security regulations required. By the mid 1940s cannabis research had effectively been eliminated.

I was not aware of the legal problems. And neither were the administration of the Weizmann Institute and even the police! In early 1963, through the administrative head of the Institute, I requested hashish for research from the police and was asked to come over to their store of confiscated smuggled material in Tel Aviv. There I drank a cup of coffee with the elderly person in charge, he told me how the police had caught hashish smugglers from the Lebanon and I told him what we wanted to explore. I received 5 kg of hashish in the form of 10 "hashish soles," signed a receipt and boarded a bus to the Weizmann Institute some 15 miles away. On the bus travelers commented on the pleasant smell of the "vegetables" I was carrying.

We had all actually broken the strict laws on illicit substances. The Ministry of Health was supposed to have approved the research, the police should not have given me such a dangerous substance and I was essentially a criminal. But at the Ministry some of the bureaucrats in charge were my ex-colleagues, and after I was severely scolded for breaking the laws, we drank together some more coffee and I got a properly signed and stamped document. Living and doing research in a small country where people working in related areas generally know each other has at times its positive aspects.

The first thing my colleague Yuval Shvo and I did with the now legal hashish was to reisolate cannabidiol (CBD)—already isolated previously by Adams and Todd—and elucidate its structure. Not surprisingly, for over 40 years after our work on CBD in 1963, very few scientists and clinicians became interested in CBD, although we showed, together with Brazilian colleagues, already in 1980, that it is a good, novel antiepileptic drug and together with Israeli and British colleagues and friends published, in the early years of the 21st century, that it is also a potent drug in autoimmune diseases. Gratifyingly, today CBD is widely acclaimed as a novel antiepileptic drug in children.

In 1964 Yehiel Gaoni, a recent PhD chemist from the Sorbonne, joined the hashish group, and together we isolated for the first time the psychoactive constituent of hashish, which we named Δ^1-tetrahydrocannabinol (Δ^1-THC). The chemical nomenclature rules demanded a change, and today Δ^1-THC is known as Δ^9-THC. However, the academic administration of the Weizmann Institute were not happy with our research. Why could not we work on more respectable scientific topics? With a heavy heart in 1966 I moved to the supposedly more conservative Hebrew University in Jerusalem, where, surprisingly, I had full support for our work and where we continued to pamper cannabinoid ideas for over 50 years.

Over the next decades many other cannabinoids were isolated, their structures were elucidated and compounds were synthesized by our group. We looked at the metabolism of the cannabinoids and we collaborated with numerous biologists in exploring the various cannabinoid activities.

Anandamide and 2-AG were discovered by my group in the 1990s, and, with colleagues in many countries, we found that the newly discovered endocannabinoid system is involved in a large number of biological reactions and clinical conditions. The pampered idea has become a scientific adult.

Looking ahead. Shall we have endocannabinoid drugs soon? By modification of the endocannabinoid system, possibly through epigenetics, can we possibly treat clinical conditions in the future? As we have previously speculated, is the subtle chemical disparity of the many dozens of endocannabinoid-like compounds in the brain somehow involved in the huge variability in personality—an area in psychology that is yet to be fully understood?

What next?

REFERENCES

1. Schultes RE, Klein WM, Plowman T, Lockwood TE. Cannabis: An example of taxonomic neglect. *Botanical Museum Leaflets, Harvard University.* 1974;23(9):337−367.

2. Merlin MD. Archaeological evidence for the tradition of psychoactive plant use in the old world. *Econ Bot.* 2003;57(3):295−323.

3. Jiang HE, Li X, Zhao YX, et al. A new insight into *Cannabis sativa* (Cannabaceae) utilization from 2500-year-old Yanghai Tombs, Xinjiang, China. *J Ethnopharmacol.* 2006;108 (3):414−422.

4. Doyle E, Spence AA. Cannabis as a medicine? *Br J Anaesth.* 1995;74(4):359−361.

5. Zuardi AW. History of cannabis as a medicine: a review. *Rev Bras Psiquiatr.* 2006;28 (2):153−157.

6. Mechoulam S, Lander N, Dikstein S, Carlini EA, Blumenthal M. On the therapeutic possibilities of some cannabinoids. In: Cohen S, Stillman R, eds. *The Therapeutic Potential of Marihuana.* New York, NY: Plenum Press; 1976:36.

7. Duke JA, Wain KK. Medicinal Plants of the world, Computer index with more than 85.000 entries. In: Duke JA, ed. *Handbook of Medicinal Herbs.* Boca Raton, FL: CRC Press; 1981:96.

8. Abrams DI, Jay CA, Shade SB, et al. Cannabis in painful HIV-associated sensory neuropathy: a randomized placebo-controlled trial. *Neurology.* 2007;68(7):515−521.

9. Pryce G, Baker D. Emerging properties of cannabinoid medicines in management of multiple sclerosis. *Trends Neurosci.* 2005;28(5):272−276.

10. Brenneisen R, Egli A, Elsohly MA, Henn V, Spiess Y. The effect of orally and rectally administered delta 9-tetrahydrocannabinol on spasticity: a pilot study with 2 patients. *Int J Clin Pharmacol Ther.* 1996;34(10):446−452.

11. Long LE, Malone DT, Taylor DA. The pharmacological actions of cannabidiol. *Drugs Future.* 2005;30(7):747.

12. Sirikantaramas S, Taura F, Morimoto S, Shoyama Y. Recent advances in *Cannabis sativa* research: biosynthetic studies and its potential in biotechnology. *Curr Pharm Biotechnol.* 2007;8(4):237−243.

13. Mechoulam R, Carlini EA. Toward drugs derived from cannabis. *Naturwissenschaften.* 1978;65(4):174–179.

14. Cunha JM, Carlini EA, Pereira AE, et al. Chronic administration of cannabidiol to healthy volunteers and epileptic patients. *Pharmacology.* 1980;21(3):175–185.

15. Ranalli P, Di Candilo M, Mandolino G, Grassi G, Carboni A. Hemp for sustainable agricultural systems. *Agro Food Ind Hi Tech.* 1999;10(2):33–38.

16. Callaway JC. Hempseed as a nutritional resource: an overview. *Euphytica.* 2004;140 (1–2):65–72.

17. Kriese U, Schumann E, Weber WE, Beyer M, Brühl L, Matthäus B. Oil content, tocopherol composition and fatty acid patterns of the seeds of 51 *Cannabis sativa* L. genotypes. *Euphytica.* 2004;137(3):339–351.

18. Small E, Marcus D. Hemp: a new crop with new uses for North America. In: Janick J, Whipkey A, eds. *Trends in New Crops and New Uses.* Alexandria, VA: ASHS Press; 2002:284–326.

19. Emboden WA. Cannabis—a polytypic genus. *Econ Bot.* 1974;28(3):304–310.

20. Hillig KW. A chemotaxonomic analysis of terpenoid variation in Cannabis. *Biochem Syst Ecol.* 2004;32(10):875–891.

21. Hillig KW. Genetic evidence for speciation in Cannabis (Cannabaceae). *Genet Resour Crop Evol.* 2005;52(2):161–180.

22. Hillig KW, Mahlberg PG. A chemotaxonomic analysis of cannabinoid variation in Cannabis (Cannabaceae). *Am J Bot.* 2004;91(6):966–975.

23. Small E. Morphological variation of achenes of Cannabis. *Can J Bot.* 1975;53(10):978–987.

24. Small E. American law and the species problem in Cannabis: science and semantics. *Bull Narc.* 1975;27(3):1–20.

25. Small E, Cronquist A. A practical and natural taxonomy for Cannabis. *Taxon.* 1976;25(4):405.

26. Gilmore S, Peakall R, Robertson J. Short tandem repeat (STR) DNA markers are hypervariable and informative in *Cannabis sativa*: implications for forensic investigations. *Forensic Sci Int.* 2003;131(1):65–74.

27. Sakamoto K, Shimomura K, Komeda Y, Kamada H, Satoh S. A male-associated DNA sequence in a dioecious plant, *Cannabis sativa* L. *Plant Cell Physiol.* 1995;36(8):1549–1554.

28. Mandolino G, Carboni A, Forapani S, Faeti V, Ranalli P. Identification of DNA markers linked to the male sex in dioecious hemp (*Cannabis sativa* L.). *Theor Appl Genet.* 1999;98 (1):86–92.

29. Flachowsky H, Schumann E, Weber WE, Peil A. Application of AFLP for the detection of sex-specific markers in hemp. *Plant Breed.* 2001;120(4):305–309.

30. Törjék O, Bucherna N, Kiss E, et al. Novel male-specific molecular markers (MADC5, MADC6) in hemp. *Euphytica.* 2002;127(2):209–218.

31. Sakamoto K, Abe T, Matsuyama T, et al. RAPD markers encoding retrotransposable elements are linked to the male sex in *Cannabis sativa* L. *Genome.* 2005;48(5):931–936.

32. Techen N, Chandra S, Lata H, ElSohly M, Khan I. Genetic identification of female *Cannabis sativa* plants at early developmental stage. *Planta Med.* 2010;76(16):1938–1939.

33. Wood TB, Spivey WTN, Easterfield TH. Charas. The resin of Indian hemp. *J Chem Soc, Trans.* 1896;69:539–546.

34. Wood TB, Spivey WTN, Easterfield TH. Cannabinol. Part I. *J Chem Soc.* 1899;75:20–36.

35. Adams R, Baker BR, Wearn RB. Structure of cannabinol. III. Synthesis of cannabinol, 1-hydroxy-3-amyl-6,6,9-trimethyl-6-dibenzopyran. *J Am Chem Soc.* 1940;62:2204–2207.

36. Cahn RS. Cannabis INDICA resin. III. Constitution of cannabinol. *J Chem Soc.* 1932;3:1342–1353.

37. Ghosh R, Todd AR, Wilkinson S. *Cannabis indica.* Part V. The synthesis of cannabinol. *J Chem Soc.* 1940;:1393–1396.

38. Adams R, Hunt M, Clark JH. Structure of cannabidiol, a product isolated from the marihuana extract of Minnesota wild hemp. I. *J Am Chem Soc.* 1940;62(1):196–200.

39. Mechoulam R, Shvo Y. Hashish—I. *Tetrahedron.* 1963;19(12):2073–2078.

40. Gaoni Y, Mechoulam R. Isolation, structure, and partial synthesis of an active constituent of hashish. *J Am Chem Soc.* 1964;86(8):1646–1647.

41. Turner CE, Elsohly MA, Boeren EG. Constituents of *Cannabis sativa* L. XVII. A review of the natural constituents. *J Nat Prod.* 1980;43(2):169–234.

42. ElSohly MA, Slade D. Chemical constituents of marijuana: the complex mixture of natural cannabinoids. *Life Sci.* 2005;78(5):539–548.

43. Mandolino G, Bagatta M, Carboni A, Ranalli P, de Meijer E. Qualitative and quantitative aspects of the inheritance of chemical phenotype in Cannabis. *J Ind Hemp.* 2003;8(2):51–72.

44. Fetterman PS, Keith ES, Waller CW, Guerrero O, Doorenbos NJ, Quimby MW. Mississippi-grown *Cannabis sativa* L.: preliminary observation on chemical definition of phenotype and variations in tetrahydrocannabinol contentversus age, sex, and plant part. *J Pharm Sci.* 1971;60(8):1246–1249.

45. Small E, Beckstead HD. Cannabinoid phenotypes in *Cannabis sativa. Nature.* 1973;245 (5421):147–148.

46. Small E, Beckstead HD. Common cannabinoid phenotypes in 350 stocks of Cannabis. *Lloydia.* 1973;36(2):144–165.

47. Fournier G, Richez-Dumanois C, Duvezin J, Mathieu JP, Paris M. Identification of a new chemotype in *Cannabis sativa*: cannabigerol-dominant plants, biogenetic and agronomic prospects. *Planta Med.* 1987;53(3):277–280.

48. Hemphill JK, Turner JC, Mahlberg PG. Cannabinoid content of individual plant organs from different geographical strains of *Cannabis sativa* L. *J Nat Prod.* 1980;43(1):112–122.

49. de Meijer EPM, van der Kamp HJ, van Eeuwijk FA. Characterisation of Cannabis accessions with regard to cannabinoid content in relation to other plant characters. *Euphytica.* 1992;62(3):187–200.

50. Chandra S, Lata H, Khan IA, Elsohly MA. Photosynthetic response of *Cannabis sativa* L. to variations in photosynthetic photon flux densities, temperature and CO_2 conditions. *Physiol Mol Biol Plants.* 2008;14(4):299–306.

51. Chandra S, Lata H, Khan IA, ElSohly MA. Variations in temperature response of photosynthesis in drug and fiber type varieties of *Cannabis sativa* L. *Planta Med.* 2009;75(04).

52. Mendoza MA, Mills DK, Lata H, Chandra S, ElSohly MA, Almirall JR. Genetic individualization of *Cannabis sativa* by a short tandem repeat multiplex system. *Anal Bioanal Chem.* 2008;393(2):719–726.

53. Bazzaz FA, Dusek D, Seigler DS, Haney AW. Photosynthesis and cannabinoid content of temperate and tropical populations of *Cannabis sativa. Biochem Syst Ecol.* 1975;3:15–18.

54. Valle JR, Vieira JE, Aucelio JG, Valio IF. Influence of photoperiodism on cannabinoid content of *Cannabis sativa* L. *Bull Narc.* 1978;30(1):67–68.

55. Bócsa I, Máthé P, Hangyel L. Effect of nitrogen on tetrahydrocannabinol (THC) content in hemp (*Cannabis sativa* L.) leaves at different positions. *J Int Hemp Assoc.* 1997;4:80–81.

56. Pate DW. Chemical ecology of Cannabis. *J Int Hemp Assoc.* 1994;1(29):32–37.

57. Shoyama Y, Hirano H, Oda M, Somehara T, Nishioka I. Cannabichromevarin and cannabigerovarin, two new propyl homologues of cannabichromene and cannabigerol. *Chem Pharm Bull (Tokyo)*. 1975;23(8):1894–1895.

58. Kajima M, Piraux M. The biogenesis of cannabinoids in *Cannabis sativa*. *Phytochemistry*. 1982;21(1):67–69.

59. Fellermeier M, Zenk MH. Prenylation of olivetolate by a hemp transferase yields cannabigerolic acid, the precursor of tetrahydrocannabinol. *FEBS Lett*. 1998;427(2):283–285.

60. Flores-Sanchez IJ, Verpoorte R. Secondary metabolism in cannabis. *Phytochem Rev*. 2008;7 (3):615–639.

61. Flores-Sanchez IJ, Verpoorte R. PKS activities and biosynthesis of cannabinoids and flavonoids in *Cannabis sativa* L. plants. *Plant Cell Physiol*. 2008;49(12):1767–1782.

62. Hartsel SC, Loh WH, Robertson LW. Biotransformation of cannabidiol to cannabielsoin by suspension cultures of *Cannabis sativa* and *Saccharum officinarum*. *Planta Med*. 1983;48 (1):17–19.

63. Mandolino G, Ranalli P. Advances in biotechnological approaches for hemp breeding and industry. In: Ranalli P, ed. *Advances in Hemp Research*. New York, NY: Haworth Press; 1999:185–208.

64. Ślusarkiewicz-Jarzina A, Ponitka A, Kaczmarek Z. Influence of cultivar, explant source and plant growth regulator on callus induction and plant regeneration of *Cannabis sativa* L. *Acta Biol Cracov Ser Bot*. 2005;47(2):145–151.

65. Lata H, Chandra S, Khan I, ElSohly MA. Thidiazuron-induced high-frequency direct shoot organogenesis of *Cannabis sativa* L. *In Vitro Cell Dev Biol-Plant*. 2009;45(1):12–19.

66. Lata H, Chandra S, Khan IA, ElSohly MA. Propagation through alginate encapsulation of axillary buds of *Cannabis sativa* L.—an important medicinal plant. *Physiol Mol Biol Plants*. 2009;15(1):79–86.

67. Lata H, Chandra S, Khan IA, ElSohly MA. Micropropagation of *Cannabis sativa*. In: Zwenger SR, ed. *Biotechnology of Cannabis sativa. 2nd ed.* New York, NY: Extreme publication Inc; 2014:82.

68. Wang R, He LS, Xia B, Tong JF, Li N, Peng F. A micropropagation system for cloning of hemp (*Cannabis sativa* L.) by shoot tip culture. *Pak J Bot*. 2009;41(2):603–608.

69. Lata H, Chandra S, Techen N, Khan IA, ElSohly MA. In vitro mass propagation of *Cannabis sativa* L.: a protocol refinement using a novel aromatic cytokinin meta-topolin and assessment of eco-physiological, biochemical and genetic fidelity of micropropagated plants (paper communicated). 2015.

70. Lata H, Bedir E, Hosick A, Ganzera M, Khan I, Moraes RM. In vitro plant regeneration from leaf-derived callus of *Cimicifuga racemosa*. *Planta Med*. 2002;68(10):912–915.

71. Fisse J, Braut F, Cosson L, Paris M. Etude in vitro des capacities organogenetiques de tissues de *Cannabis sativa* L. Effet de differentes substances de crossance. *Planta Med*. 1981;15:217–223.

72. Lata H, Chandra S, Khan IA, Elsohly MA. High frequency plant regeneration from leaf derived callus of high Delta9-tetrahydrocannabinol yielding *Cannabis sativa* L. *Planta Med*. 2010;76(14):1629–1633.

73. Lata H, Chandra S, Techen N, Khan IA, ElSohly MA. Assessment of the genetic stability of micropropagated plants of *Cannabis sativa* by ISSR markers. *Planta Med*. 2010;76 (1):97–100.

74. Lata H, Chandra S, Techen N, Wang Y-H, Khan IA. Molecular analysis of genetic fidelity in micropropagated plants of *Stevia rebaudiana* Bert. Using ISSR marker. *AJPS*. 2013;04 (05):964–971.

75. Chandra S, Lata H, Mehmedic Z, Khan IA, ElSohly MA. Assessment of cannabinoids content in micropropagated plants of *Cannabis sativa* and their comparison with conventionally propagated plants and mother plant during developmental stages of growth. *Planta Med.* 2010;76(7):743–750.

76. Holden JHW, Williams JT, eds. *Crop Genetic Resources: Conservation and Evaluation.* London: George Allen & Unwin (Publishers) Ltd; 1984.

77. Lata H, Chandra S, Mehmedic Z, Khan IA, ElSohly MA. In vitro germplasm conservation of high Δ^9-tetrahydrocannabinol yielding elite clones of *Cannabis sativa* L. under slow growth conditions. *Acta Physiol Plant.* 2012;34(2):743–750.

78. Chandra S, Lata H, Mehmedic Z, Khan IA, ElSohly MA. Light dependence of photosynthesis and water vapor exchange characteristics in different high Δ^9-THC yielding varieties of *Cannabis sativa* L. *J Appl Res Med Aromat Plants.* 2015;2(2):39–47.

79. Chandra S, Lata H, Khan IA, ElSohly MA. Temperature response of photosynthesis in different drug and fiber varieties of *Cannabis sativa* L. *Physiol Mol Biol Plants.* 2011;17(3):297–303.

80. Kimball BA. Carbon dioxide and agricultural yield: an assemblage and analysis of 430 prior observations. *Agron J.* 1983;75(5):779.

81. Kimball BA. *Carbon dioxide and agricultural yield: an assemblage and analysis of 770 prior observations. Water Conservation Lab Report 14, US Water Conservation Lab.* Phoenix, AZ: USDA-ARS; 1983.

82. Kimball BA. Influence of elevated CO_2 on crop yield. Physiology Yield and Economics In: Enoch HZ, Kimball BA, eds. *Carbon Dioxide Enrichment of Greenhouse Crops.* vol. 2. Boca Raton, FL: CRC Press, Inc; 1986:105–115.

83. Poorter H. Interspecific variation in the growth response of plants to an elevated ambient CO_2 concentration. *Vegetatio.* 1993;104–105(1):77–97.

84. Idso K. Plant responses to atmospheric CO_2 enrichment in the face of environmental constraints: a review of the past 10 years' research. *Agric For Meteorol.* 1994;69(3–4):153–203.

85. Chandra S, Lata H, Khan IA, ElSohly MA. Photosynthetic response of *Cannabis sativa* L., an important medicinal plant, to elevated levels of CO_2. *Physiol Mol Biol Plants.* 2011;17 (3):291–295.

86. Narayanaswami K, Golani HC, Bami HL, Dau RD. Stability of *Cannabis sativa* L. samples and their extracts, on prolonged storage in Delhi. *Bull Narc.* 1978;30(4):57–69.

87. Harvey DJ. Stability of cannabinoids in dried samples of cannabis dating from around 1896–1905. *J Ethnopharmacol.* 1990;28(1):117–128.

88. Mehmedic Z, Chandra S, Slade D, et al. Potency trends of Delta9-THC and other cannabinoids in confiscated cannabis preparations from 1993 to 2008. *J Forensic Sci.* 2010;55 (5):1209–1217.

CHAPTER 2

Biosynthesis and Pharmacology of Phytocannabinoids and Related Chemical Constituents

In addition to nucleic acids, proteins, lipids, and carbohydrates, cannabis produces a large number of additional constituents, or secondary metabolites, including phytocannabinoids, terpenoids, and phenylpropanoids.[1,2] While phytocannabinoids are often referred to as the "active" ingredients in cannabis, these other chemical constituents have a broad spectrum of pharmacological properties and can contribute to the effects seen upon cannabis ingestion or combustion and inhalation, and may also be contained within and contribute to the activity of extracts, tinctures, and other cannabis formulations.[3] This overview of cannabis constituents will focus on the phytocannabinoids, terpenoids, and flavonoids that make up a large percentage of the pharmacologically active ingredients of current or emerging interest.

PHYTOCANNABINOID CONSTITUENTS IN CANNABIS

Phytocannabinoids are a structurally diverse class of naturally occurring chemical constituents in the genus *Cannabis* (Cannabaceae). This chemical classification is broadly based on their derivation from a common C21 precursor (cannabigerolic acid,[4] CBGA), or its C19 analog (cannabigerovaric acid,[5] CBGVA), the predominate phytocannabinoid precursors formed through the reaction of geranyl pyrophosphate with olivetolic and divarinic acid, respectively (Fig. 2.1).

Enzymatic conversion of cannabigerolic and cannabidivaric acid produces a wide variety of C21 terpenophenolics,[6] including (−)-*trans*-Δ^9-tetrahydrocannabinol (Δ^9-THC), cannabigerol (CBG), cannabichromene (CBC), cannabicyclol (CBL), cannabidiol (CBD), cannabinodiol (CBND), and cannabinol (CBN), and their C19

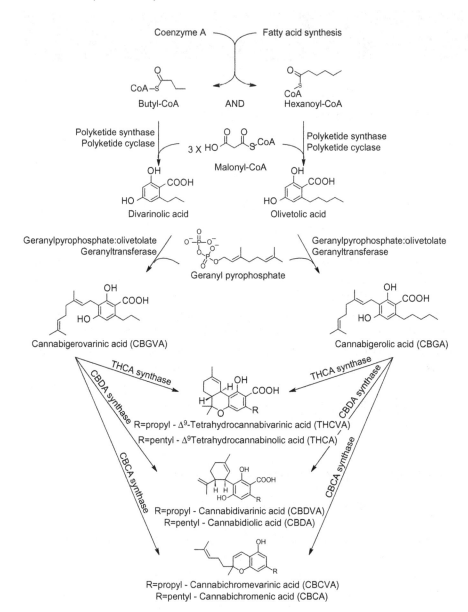

Figure 2.1 Biosynthesis of phytocannabinoids.

homologs Δ^9-tetrahydrocannabivarin (Δ^9-THCV), cannabivarin (CBV), and cannabidivarin (CBDV). More than 100 phytocannabinoids across 11 chemical classes have been isolated and identified to date.[7] In the growing *Cannabis sativa* plant, most of these cannabinoids are initially formed as carboxylic acids (eg, Δ^9-THCA, CBDA, CBCA, and

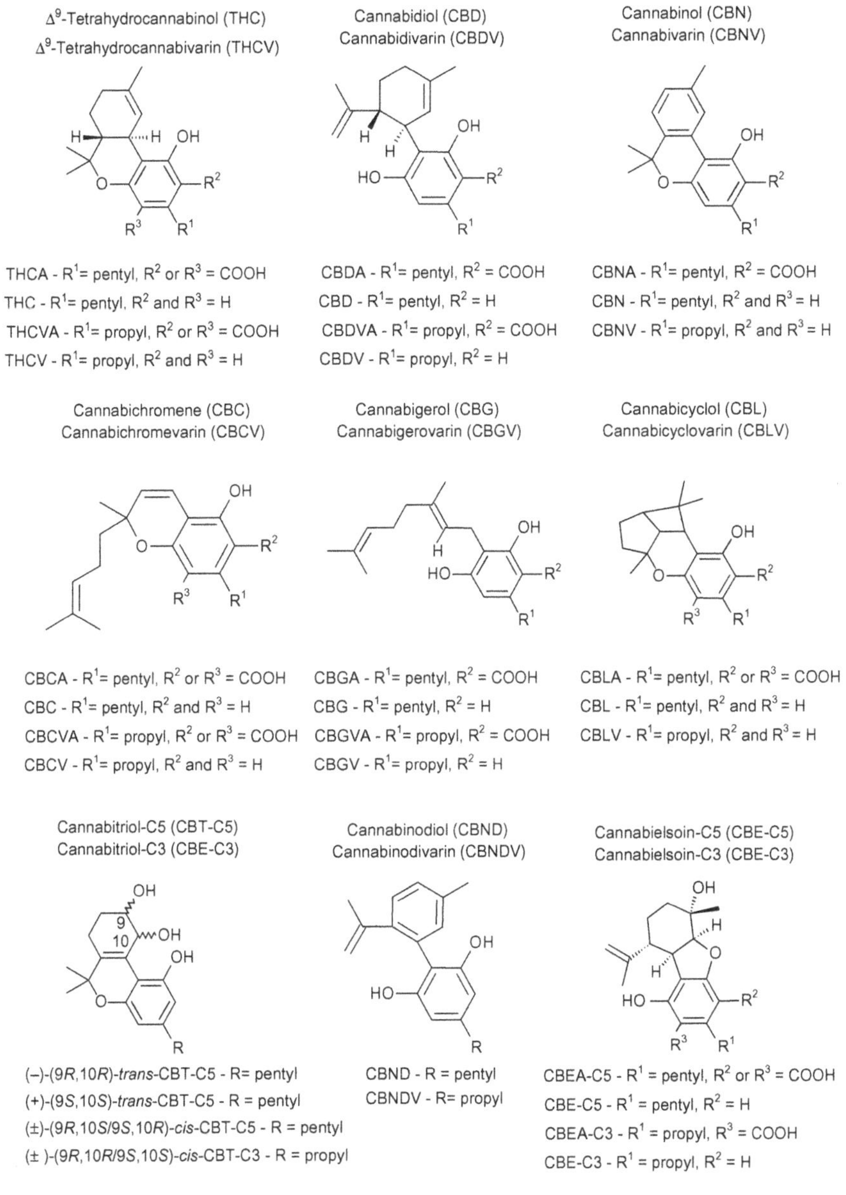

Figure 2.2 Primary phytocannabinoid constituents in cannabis.

Δ^9-THCVA) that are decarboxylated to their corresponding neutral forms as a consequence of drying, heating, combustion, or aging (Fig. 2.2). There are also different isomers of phytocannabinoids resulting from variations or isomerization in the position of the double bond in the alicyclic carbon ring (eg, $(-)$-*trans*-Δ^8-THC). It is important to note that CBN is not formed biosynthetically, but is an oxidative degradant of Δ^9-THC.[2]

Regulation of cannabinoid content in each plant phenotype (chemotype) has been proposed to involve genetic control of the expression of a variety of synthetic enzymes by four independent loci.[8] Qualitatively, the cannabinoid chemotype is controlled by the variation in expression of these phytocannabinoid synthetic enzymes, resulting in progenies and populations that have discrete distributions of chemical composition (ie, chemical ratios of phytocannabinoids, such as the Δ^9-THC/CBD ratio). Quantitatively, the phytocannabinoid content is controlled by polygenic mechanisms, and is strongly influenced by environmental factors, such that a Gaussian distribution of total cannabinoid content is typically observed. In addition, the cannabinoid content and profile changes over time as the plant grows, matures, and ages.[9] Wild-type chemotypes can therefore differ between Δ^9-THCA predominance and CBDA predominance in discrete populations, but vary dramatically in total cannabinoid content, with clones of both types reaching total cannabinoid content levels of up to 25−30% (w/w) of the dry and trimmed inflorescences. Spontaneous mutations and selective breeding have produced unique chemotypes that show CBGA-,[10] CBCA-,[11] or Δ^9-THCVA predominance,[8] as well as cannabinoid-free chemotypes.[12] Selective breeding has produced hundreds of strains that differ in appearance and chemical composition, and patients and recreational users often prefer specific strains for their purported ability to produce specific pharmacological effects. De Meijer speculates that future breeding might produce novel terpenophenolic compounds such as those with branched alkyl or aromatic side chains, or chemotypes with increased ratios of currently minor constituents such as methyl, butyl, or farnesyl cannabinoids.[8]

The current variation in phytocannabinoid content across and within chemotypes has important implications in medicinal cannabis and cannabis-based formulations and dosing. This has become increasingly apparent and can be recognized by the plethora of varieties of cannabis being cultivated, manufactured, and marketed as dosing formulations in the medicinal and recreational market. Similarly, the nonphytocannabinoid composition of cannabis is receiving increasing pharmacological attention, particularly terpenoids and flavonoids.[13]

MONOTERPENOID, SESQUITERPENOID, AND DITERPENOID CONSTITUENTS OF CANNABIS

Geranyl pyrophosphate is the precursor in the synthesis of the more ubiquitous terpenoids (Fig. 2.3), leading to the formation of limonene and other monoterpenoids in secretory cell plastids,[14] or coupling with isopentenyl pyrophosphate in the cytoplasm to form farnesyl pyrophosphate, a key intermediate in the biosynthesis of sesquiterpenoids and triterpenoids.[13,15] Addition of another isopentenyl group to farnesyl pyrophosphate leads to geranylgeranyl pyrophosphate, which is the precursor of the diterpenoids.[15] Some of the predominate volatile monoterpenes found in cannabis are β-myrcene, (E)-β-ocimene, terpinolene, limonene, and β-pinene.[16,17] β-caryophyllene, α-caryophyllene

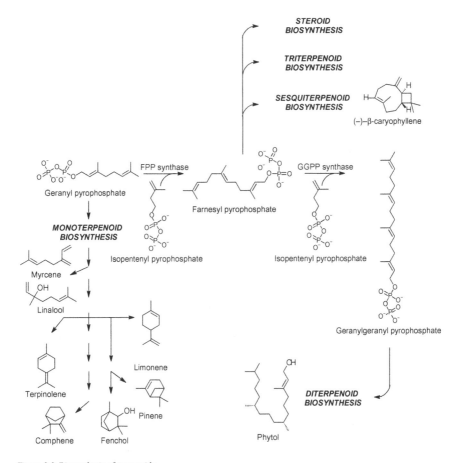

Figure 2.3 Biosynthesis of terpenoids.

(humulene), longifolene, α-zingiberene, and β-cedrene are among the major sesquiterpenes found in prototypical cannabis samples.[16]

PHENYLPROPANOID CONSTITUENTS OF CANNABIS

Phenylpropanoids are of interest for their diverse pharmacological and industrial applications.[18] Their biosynthesis begins with phenylalanine, derived from the shikimate pathway, which is converted by phenylalanine ammonia lyase into cinnamic acid (Fig. 2.4). After hydroxylation of cinnamic acid by cinnamate-4-hydroxylase to form *p*-coumaric acid, it is converted in *p*-coumaroyl CoA by addition of a CoA thioester by a 4-coumarate:CoA ligase enzyme. This common high energy intermediate is used in the biosynthesis of cell wall constituents (lignins), pigments (flavonoids, antocians), and UV protectant and pest resistance compounds (stilbenoids, flavonoids, isoflavonoids, coumarins, and furanocoumarins). A key enzyme in the flavonoid biosynthesis pathway is chalcone synthase (CHS), a protein in the superfamily of polyketide synthase that includes stilbene synthase, phlorisovalerophenone synthase, isobutyrophenone synthase (BUS), and olivetol synthase activities that can be detected during the development and growth of glandular trichomes on bracts of cannabis.[18,19] The activities of polyketide synthases and the resulting biosynthesis of cannabinoids, stilbenoids, and flavonoids in the plant are induced in response to a wide range of stimuli such as UV light, pathogens, hormones, elicitors, growth substances, and wounding. Some of the naturally occurring flavonoid constituents are orientin, vitesin, luteolin-7-*O*-β-D-glucuronide, and apigenin-7-*O*-β-D-glucuronide.[20]

THERAPEUTIC INDICATIONS FOR MEDICINAL CANNABIS AND CANNABIS-DERIVED DOSAGE FORMULATIONS

Cannabis and cannabis-derived dosage formulations such as hashish have a long history of medicinal use. However, the intoxicating effects and associated abuse liability, scheduling, and control efforts have limited the number of methodologically rigorous clinical studies. The most widely supported indications for herbal cannabis and cannabis-derived medicines are nausea and vomiting in cancer chemotherapy, anorexia and cachexia in HIV/AIDS, chronic and neuropathic pain, and spasticity in multiple sclerosis and spinal cord injury.[21] A recent review of the randomized clinical trials investigating cannabis and

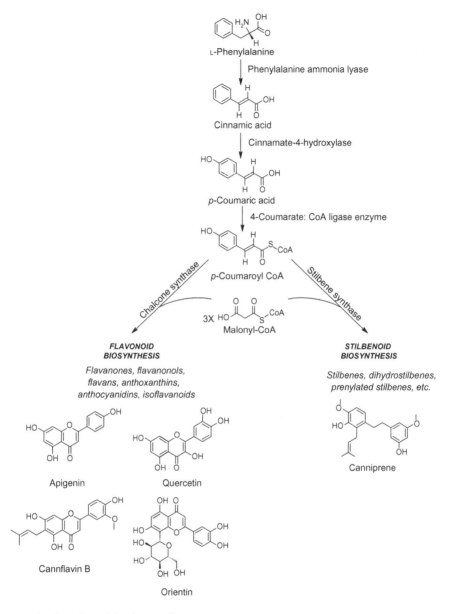

Figure 2.4 Biosynthesis of phenylpropanoids.

cannabinoid therapeutics provides some support for these indications as well as for sleep disorders and Tourette's syndrome.[22] Glaucoma was one of the more frequently cited medical indications in the late twentieth century; however, this indication has received diminished support from the medical community due to the success of surgical approaches and the availability of drugs with greater efficacy.[21] Cannabis also has a long history of use to control epileptic seizures, the subject of recent reviews.[23] The use of cannabis or cannabinoid-derived drugs for seizures is not fully supported by the Institute of Medicine; however, CBD has gained considerable attention and anecdotal support as a treatment for Dravet Syndrome and other specific seizure disorders.[24] Other indications being investigated are irritable bowel syndrome[25–28] and posttraumatic stress disorder.[29,30]

PHARMACOLOGICAL EFFECTS OF CANNABIS CONSTITUENTS

The contribution of the various chemical constituents in cannabis to its therapeutic and organoleptic effects varies because of several factors, including their differing concentrations (content), chemical properties (eg, stability, volatility), pharmacological actions (eg, receptor affinities, efficacies), physicochemical parameters (eg, lipophilicity, solubility), pharmacokinetics, and pharmacodynamics. The principal acidic forms of phytocannabinoids produced by plant biosynthesis have generally been considered devoid of psychoactivity in man and laboratory animals.[31] Δ^9-THCA, CBNA, and CBGA have previously been reported to lack affinity at the CB1 receptor.[32] More recently, however, Δ^9-THCA showed antinausea and antiemetic effects in animals, with greater potency than Δ^9-THC,[33] and the effect was reversed by rimonabant, the CB1 receptor antagonist. Δ^9-THCA was recently reported to bind to CB1 and CB2 receptors with greater affinity than Δ^9-THC.[34] Δ^9-THCA has shown immunomodulatory actions that are CB1- and CB2 receptor–independent.[35] Thus, further studies of the pharmacological activity and mechanism of action of Δ^9-THCA are needed.

The primary psychoactive phytocannabinoid is often considered to be Δ^9-THC because of its rapid formation from Δ^9-THCA through decarboxylation during combustion of plant material resulting in its high concentration in smoke during inhalation and its potency at cannabinoid receptors at physiologically relevant concentrations. In support of this hypothesis, radioligand binding studies have shown that Δ^9-THC binds to the CB1 receptor with high affinity (K_i values ~ 50 nM), while CBN has an approximately 10-fold lower affinity, and CBD and CBG have K_i values estimated to be greater than

500 nM.[36,37] This receptor binding affinity correlates to both the inhibition of adenylate cyclase in vitro and the analgesic and psychoactivity of these compounds in vivo.[38] Δ^9-THC also binds to the CB2 receptor with similar high affinity and inhibits adenylate cyclase.[39] However, it is important to note that Δ^9-THC acts as a partial agonist in GTP-γ-S assays[40] as compared to the effects seen with more efficacious synthetic cannabinoid agonists such as CP-55940 and WIN55212-2.[41] Because it acts as a partial agonist at both CB1 and CB2, it elicits a response that is strongly influenced by the tissue-specific expression levels, their extent of constitutive signaling, and the ongoing endogenous cannabinoid release or tone of the receptor system.[42] In addition to its production of psychoactivity in man, Δ^9-THC produces a myriad of additional cannabinoid receptor–mediated pharmacological effects in laboratory animals and man.[43]

Several other phytocannabinoids bind to and modulate cannabinoid receptor function. Δ^8-THC, for example, binds to CB1 and CB2 with affinities approximating those of Δ^9-THC and acts as an agonist.[44–46] Δ^9-THCV also binds to CB1 and CB2 with nanomolar affinities[47] and acts as a cannabimimetic agonist. High doses of Δ^9-THCV have been reported to produce a psychoactive effect characterized as mild intoxication in man[31,48] and catalepsy[49] and analgesia[50] in laboratory animals. However, Δ^9-THCV can also antagonize cannabinoid receptor agonists in CB1-expressing tissues in a manner that is both tissue- and ligand-dependent.[42,50] Moreover, this compound is reported to be capable of behaving either as a CB1 antagonist or, at higher doses, as a CB1 agonist in vivo. CBN can also bind to CB1[51] and CB2, but does so with lower affinity than Δ^9-THC. It also fails to inhibit forskolin-stimulated cAMP increase at doses up to 1 μM in transfected cell lines expressing CB1 or CB2 receptors,[52] and produces only mild intoxication in man at high intravenous doses.[31] Other than Δ^9-THC, Δ^8-THC, CBN, and Δ^9-THCV, no phytocannabinoids have been reported to activate CB1 or CB2 receptors with nanomolar or low micromolar potency.[53] Neither CBD nor CBDA is psychoactive in man,[31] nor does CBD bind to the CB1[51] or CB2[54] receptor with high affinity. However, Pertwee has reported that CBD displays high potency as an antagonist in CB1- and CB2-expressing cells or tissues.[42]

The phytocannabinoids in cannabis can act via a plethora of other noncannabinoid receptor–mediated systems to produce a wide variety of additional pharmacological effects. As reviewed by Pertwee in 2010, Δ^9-THC and other phytocannabinoids have been shown to modulate the activity of GPR-55[55] and many other receptors and enzyme systems.[56] For instance, CBD, CBN, and Δ^9-THC produce noncompetitive/allosteric interactions with μ and δ opioid receptors,[57,58] and Δ^9-THC allosterically modulates glycine receptor activation at nanomolar concentrations.[59] Δ^9-THC and CBD inhibit Ca(V)3 channels at pharmacologically relevant concentrations, and Δ^9-THC but not CBD may increase the amount of calcium entry following T-type channel activation by stabilizing open states of the channel.[60] These phytocannabinoids also interact with transient receptor potential (TRP) channels and enzymes of the endocannabinoid system.[61] De Petrocellis et al. showed that both CBDV and CBD activate and desensitize transient receptor potential vanilloid 1 (TRPV1) channels, and that CBC disrupts transport of the endocannabinoid anandamide in vitro.[62] CBG, CBGV, and Δ^9-THCV stimulate and desensitize human TRPV1, while Δ^9-THC, CBC, CBD, and CBN are potent rat TRPA1 agonists and desensitizers. All of these cannabinoids except CBC and CBN also potently activate and desensitize rat TRPV2.[62] Interestingly, in three models of seizure, cannabis-derived "botanical drug substances" rich in CBDV and CBD exerted significant anticonvulsant effects that were not mediated by the CB1 receptor and were of comparable efficacy with purified CBDV.[63] Whether the anticonvulsant activity produced by certain phytocannabinoids and cannabis-derived drug formulations in vivo is related to these effects on calcium channels remains to be determined; however, these actions are indicative of potential therapeutic utility in the treatment of neuronal hyperexcitability. Both Δ^9-THC and CBN induce a cannabinoid receptor– independent release of calcitonin gene-related peptide (CGRP) from capsaicin-sensitive perivascular sensory nerves.[64] Thus, the antinociceptive actions of phytocannabinoids may rely on the activation of inhibitory cannabinoid receptors (CB1) in the peripheral and central nervous systems, as well as on the activation of excitatory ionotropic TRP channels coexpressed with CB1 in primary nociceptive neurons that contain and release CGRP upon activation.[65] Δ^9-THC can also act as an agonist at the peroxisome proliferator-activated receptor, whereas Δ^9-THCV does not. Thus it appears that these highly lipophilic phytocannabinoids[66] can gain access to and modulate a variety of non-CB1, non-CB2 G protein–coupled receptors, transmitter-gated channels, ion channels, and/or nuclear receptors.[56]

Nonphytocannabinoid terpenoids in cannabis contribute to the organoleptic properties of the plant, but can also modulate the activity of cannabinoid receptors and contribute to a wide variety of noncannabinoid receptor–mediated pharmacological effects.[13] For example, the sesquiterpenoid β-caryophyllene has been shown to bind to CB2 with nanomolar affinity.[67] Upon binding to CB2, it acts like a prototypical CB2 agonist and inhibits adenylate cylcase, produces intracellular calcium transients, and activates the mitogen-activated protein kinase Erk1/2 and p38 pathways in primary human monocytes. β-Caryophyllene reduces the carrageenan-induced inflammatory response in wild-type mice but not in mice lacking CB2 receptors, evidence that this terpenoid exerts cannabimimetic effects in vivo. In addition to modulation of cannabinoid receptors, terpenoids and phenylpropanoids in cannabis have potent antioxidant,[68] anticancer,[69,70] and antiinflammatory activity.[71–75] Several monoterpenic alcohols, including geraniol, nerol, and citronellol, have been reported to be promiscuous TRP modulators.[76–78] It has been hypothesized that the broad spectrum and prolonged sensory inhibition produced by phytocannabinoids and terpenoids may allow them to act synergistically as therapeutics for allodynia, itch, and other types of pain involving superficial sensory nerves and skin.[78–80] Russo has proposed therapeutic synergies and interactions among phytocannabinoids, terpenoids, and phenylpropanoids. If clinically proven, this increases the likelihood of an extensive pipeline of new therapeutic products and cannabis-derived botanical drug products.[13]

REFERENCES

1. ElSohly MA, Slade D. Chemical constituents of marijuana: the complex mixture of natural cannabinoids. *Life Sci.* 2005;78(5):539–548.

2. Turner CE, Elsohly MA, Boeren EG. Constituents of *Cannabis sativa* L. XVII. A review of the natural constituents. *J Nat Prod.* 1980;43(2):169–234.

3. McPartland JM, Russo EB. Cannabis and cannabis extracts: greater than the sum of their parts? *J Cannabis Ther.* 2001;3/4:103–132.

4. Yamauchi T, Shoyama Y, Matsuo Y, Nishioka I. Cannabigerol monomethyl ether, a new component of hemp. *Chem Pharm Bull (Tokyo).* 1968;16(6):1164–1165.

5. Shoyama Y, Hirano H, Makino H, Umekita N, Nishioka I. Cannabis. 10. Isolation and structures of 4 new propyl cannabinoid acids, tetrahydrocannabivarinic acid, cannabidivarinic acid, cannabichromevarinic acid, and cannabigerovarinic acid, from Thai Cannabis, MEAO variant. *Chem Pharm Bull.* 1977;25(9):2306–2311.

6. Shoyama Y, Yagi M, Nishioka I, Yamauchi T. Cannabis 8. Biosynthesis of cannabinoid acids. *Phytochemistry.* 1975;14(10):2189–2192.

7. ElSohly MA, Gul W. Constituents of *Cannabis sativa*. In: Pertwee RG, ed. *Handbook of Cannabis*. New York, NY: Oxford University Press; 2014:3–22.

8. de Meijer EP. The chemical phenotypes (chemotypes) of Cannabis. In: Pertwee RG, ed. *Handbook of Cannabis*. New York, NY: Oxford University Press; 2014:89–110.

9. Pacifico D, Miselli F, Carboni A, Moschella A, Mandolino G. Time course of cannabinoid accumulation and chemotype development during the growth of *Cannabis sativa* L. *Euphytica*. 2008;160(2):231–240.

10. de Meijer EPM, Hammond KM. The inheritance of chemical phenotype in *Cannabis sativa* L. (II): Cannabigerol predominant plants. *Euphytica*. 2005;145(1–2):189–198.

11. de Meijer EPM, Hammond KM, Micheler M. The inheritance of chemical phenotype in *Cannabis sativa* L. (III): variation in cannabichromene proportion. *Euphytica*. 2009;165 (2):293–311.

12. de Meijer EPM, Hammond KM, Sutton A. The inheritance of chemical phenotype in *Cannabis sativa* L. (IV): cannabinoid-free plants. *Euphytica*. 2009;168(1):95–112.

13. Russo EB. Taming THC: potential cannabis synergy and phytocannabinoid-terpenoid entourage effects. *Br J Pharmacol*. 2011;163(7):1344–1364.

14. Loza-Tavera H. Monoterpenes in essential oils. Biosynthesis and properties. *Adv Exp Med Biol*. 1999;464:49–62.

15. Dewick PM. The biosynthesis of C5-C25 terpenoid compounds. *Nat Prod Rep*. 2002;19 (2):181–222.

16. Rothschild M, Bergstrom G, Wangberg SA. *Cannabis sativa*: volatile compounds from pollen and entire male and female plants of two variants, Northern Lights and Hawaian Indica. *Bot J Linn Soc*. 2005;147(4):387–397.

17. Ross SA, ElSohly MA. The volatile oil composition of fresh and air-dried buds of *Cannabis sativa*. *J Nat Prod*. 1996;59(1):49–51.

18. Docimo T, Consonni R, Coraggio I, Mattana M. Early phenylpropanoid biosynthetic steps in *Cannabis sativa*: link between genes and metabolites. *Int J Mol Sci*. 2013;14(7):13626–13644.

19. Flores-Sanchez IJ, Verpoorte R. PKS activities and biosynthesis of cannabinoids and flavonoids in *Cannabis sativa* L. plants. *Plant Cell Physiol*. 2008;49(12):1767–1782.

20. Vanhoenacker G, Van Rompaey P, De Keukeleire D, Sandra P. Chemotaxonomic features associated with flavonoids of cannabinoid-free cannabis (*Cannabis sativa* subsp. sativa L.) in relation to hops (*Humulus lupulus* L.). *Nat Prod Lett*. 2002;16(1):57–63.

21. Institute of Medicine. Washington DC: National Academy of Sciences; 1999.

22. Whiting PF, Wolff RF, Deshpande S, et al. Cannabinoids for medical use: a systematic review and meta-analysis. *JAMA*. 2015;313(24):2456–2473.

23. Gloss D, Vickrey B. Cannabinoids for epilepsy. *Cochrane Database Syst Rev*. 2014;3: CD009270.

24. Hussain SA, Zhou R, Jacobson C, et al. Perceived efficacy of cannabidiol-enriched cannabis extracts for treatment of pediatric epilepsy: a potential role for infantile spasms and Lennox-Gastaut syndrome. *Epilepsy Behav*. 2015;47:138–141.

25. Esfandyari T, Camilleri M, Ferber I, Burton D, Baxter K, Zinsmeister AR. Effect of a cannabinoid agonist on gastrointestinal transit and postprandial satiation in healthy human subjects: a randomized, placebo-controlled study. *Neurogastroenterol Motil*. 2006;18(9):831–838.

26. Esfandyari T, Camilleri M, Busciglio I, Burton D, Baxter K, Zinsmeister AR. Effects of a cannabinoid receptor agonist on colonic motor and sensory functions in humans: a randomized, placebo-controlled study. *Am J Physiol Gastrointest Liver Physiol*. 2007;293(1): G137–145.

27. Storr M, Devlin S, Kaplan GG, Panaccione R, Andrews CN. Cannabis use provides symptom relief in patients with inflammatory bowel disease but is associated with worse disease prognosis in patients with Crohn's disease. *Inflamm Bowel Dis.* 2014;20(3):472−480.

28. Naftali T, Mechulam R, Lev LB, Konikoff FM. Cannabis for inflammatory bowel disease. *Dig Dis.* 2014;32(4):468−474.

29. Trezza V, Campolongo P. The endocannabinoid system as a possible target to treat both the cognitive and emotional features of post-traumatic stress disorder (PTSD). *Front Behav Neurosci.* 2013;7:100.

30. Betthauser K, Pilz J, Vollmer LE. Use and effects of cannabinoids in military veterans with posttraumatic stress disorder. *Am J Health Syst Pharm.* 2015;72(15):1279−1284.

31. Razdan RK. Structure-activity relationships in cannabinoids. *Pharmacol Rev.* 1986;38 (2):75−149.

32. Ahmed SA, Ross SA, Slade D, et al. Cannabinoid ester constituents from high-potency *Cannabis sativa. J Nat Prod.* 2008;71(4):536−542.

33. Rock EM, Kopstick RL, Limebeer CL, Parker LA. Tetrahydrocannabinolic acid reduces nausea-induced conditioned gaping in rats and vomiting in *Suncus murinus. Br J Pharmacol.* 2013;170(3):641−648.

34. Rosenthaler S, Pohn B, Kolmanz C, et al. Differences in receptor binding affinity of several phytocannabinoids do not explain their effects on neural cell cultures. *Neurotoxicol Teratol.* 2014;46:49−56.

35. Verhoeckx KC, Korthout HA, van Meeteren-Kreikamp AP, et al. Unheated *Cannabis sativa* extracts and its major compound THC-acid have potential immuno-modulating properties not mediated by CB1 and CB2 receptor coupled pathways. *Int Immunopharmacol.* 2006;6 (4):656−665.

36. Howlett AC, Barth F, Bonner TI, et al. International union of pharmacology. XXVII. Classification of cannabinoid receptors. *Pharmacol Rev.* 2002;54(2):161−202.

37. Devane WA, Dysarz III FA, Johnson MR, Melvin LS, Howlett AC. Determination and characterization of a cannabinoid receptor in rat brain. *Mol Pharmacol.* 1988;34(5):605−613.

38. Howlett AC. Cannabinoid inhibition of adenylate cyclase: relative activity of constituents and metabolites of marihuana. *Neuropharmacology.* 1987;26(5):507−512.

39. Slipetz DM, O'Neill GP, Favreau L, et al. Activation of the human peripheral cannabinoid receptor results in inhibition of adenylyl cyclase. *Mol Pharmacol.* 1995;48(2):352−361.

40. Sim LJ, Hampson RE, Deadwyler SA, Childers SR. Effects of chronic treatment with delta9-tetrahydrocannabinol on cannabinoid-stimulated [35S]GTPgammaS autoradiography in rat brain. *J Neurosci.* 1996;16(24):8057−8066.

41. Breivogel CS, Selley DE, Childers SR. Cannabinoid receptor agonist efficacy for stimulating [35S]GTPγS binding to rat cerebellar membranes correlates with agonist-induced decreases in GDP affinity. *J Biol Chem.* 1998;273(27):16865−16873.

42. Pertwee RG. The diverse CB1 and CB2 receptor pharmacology of three plant cannabinoids: delta9-tetrahydrocannabinol, cannabidiol and delta9-tetrahydrocannabivarin. *Br J Pharmacol.* 2008;153(2):199−215.

43. Adams IB, Martin BR. Cannabis: pharmacology and toxicology in animals and humans. *Addiction.* 1996;91(11):1585−1614.

44. Felder CC, Veluz JS, Williams HL, Briley EM, Matsuda LA. Cannabinoid agonists stimulate both receptor- and non-receptor-mediated signal transduction pathways in cells transfected with and expressing cannabinoid receptor clones. *Mol Pharmacol.* 1992;42(5):838−845.

45. Matsuda LA, Lolait SJ, Brownstein MJ, Young AC, Bonner TI. Structure of a cannabinoid receptor and functional expression of the cloned cDNA. *Nature.* 1990;346(6284):561−564.

46. Gerard CM, Mollereau C, Vassart G, Parmentier M. Molecular cloning of a human cannabinoid receptor which is also expressed in testis. *Biochem J*. 1991;279(Pt 1):129−134.

47. Thomas A, Stevenson LA, Wease KN, et al. Evidence that the plant cannabinoid Delta9-tetrahydrocannabivarin is a cannabinoid CB1 and CB2 receptor antagonist. *Br J Pharmacol*. 2005;146(7):917−926.

48. Hollister LE. Structure-activity relationships in man of cannabis constituents, and homologs and metabolites of delta9-tetrahydrocannabinol. *Pharmacology*. 1974;11(1):3−11.

49. Gill EW, Paton WD, Pertwee RG. Preliminary experiments on the chemistry and pharmacology of cannabis. *Nature*. 1970;228(5267):134−136.

50. Pertwee RG, Thomas A, Stevenson LA, et al. The psychoactive plant cannabinoid, Delta9-tetrahydrocannabinol, is antagonized by Delta8- and Delta9-tetrahydrocannabivarin in mice in vivo. *Br J Pharmacol*. 2007;150(5):586−594.

51. Thomas BF, Gilliam AF, Burch DF, Roche MJ, Seltzman HH. Comparative receptor binding analyses of cannabinoid agonists and antagonists. *J Pharmacol Exp Ther*. 1998;285(1):285−292.

52. Felder CC, Joyce KE, Briley EM, et al. Comparison of the pharmacology and signal transduction of the human cannabinoid CB1 and CB2 receptors. *Mol Pharmacol*. 1995;48(3):443−450.

53. Cascio MG, Pertwee RG. Known pharmacological actions of nine nonpsychotropic phytocannabinoids. In: Pertwee RG, ed. *Handbook of Cannabis*. New York, NY: Oxford University Press; 2014:137−156.

54. Showalter VM, Compton DR, Martin BR, Abood ME. Evaluation of binding in a transfected cell line expressing a peripheral cannabinoid receptor (CB2): identification of cannabinoid receptor subtype selective ligands. *J Pharmacol Exp Ther*. 1996;278(3):989−999.

55. Ryberg E, Larsson N, Sjogren S, et al. The orphan receptor GPR55 is a novel cannabinoid receptor. *Br J Pharmacol*. 2007;152(7):1092−1101.

56. Pertwee RG. Receptors and channels targeted by synthetic cannabinoid receptor agonists and antagonists. *Curr Med Chem*. 2010;17(14):1360−1381.

57. Vaysse PJ, Gardner EL, Zukin RS. Modulation of rat brain opioid receptors by cannabinoids. *J Pharmacol Exp Ther*. 1987;241(2):534−539.

58. Kathmann M, Flau K, Redmer A, Trankle C, Schlicker E. Cannabidiol is an allosteric modulator at mu- and delta-opioid receptors. *Naunyn-Schmiedeberg's Arch Pharmacol*. 2006;372(5):354−361.

59. Hejazi N, Zhou C, Oz M, Sun H, Ye JH, Zhang L. Delta9-tetrahydrocannabinol and endogenous cannabinoid anandamide directly potentiate the function of glycine receptors. *Mol Pharmacol*. 2006;69(3):991−997.

60. Ross HR, Napier I, Connor M. Inhibition of recombinant human T-type calcium channels by Delta9-tetrahydrocannabinol and cannabidiol. *J Biol Chem*. 2008;283(23):16124−16134.

61. De Petrocellis L, Ligresti A, Moriello AS, et al. Effects of cannabinoids and cannabinoid-enriched Cannabis extracts on TRP channels and endocannabinoid metabolic enzymes. *Br J Pharmacol*. 2011;163(7):1479−1494.

62. Iannotti FA, Hill CL, Leo A, et al. Nonpsychotropic plant cannabinoids, cannabidivarin (CBDV) and cannabidiol (CBD), activate and desensitize transient receptor potential vanilloid 1 (TRPV1) channels in vitro: potential for the treatment of neuronal hyperexcitability. *ACS Chem Neurosci*. 2014;5(11):1131−1141.

63. Hill TD, Cascio MG, Romano B, et al. Cannabidivarin-rich cannabis extracts are anticonvulsant in mouse and rat via a CB1 receptor-independent mechanism. *Br J Pharmacol*. 2013;170(3):679−692.

64. Zygmunt PM, Andersson DA, Hogestatt ED. Delta 9-tetrahydrocannabinol and cannabinol activate capsaicin-sensitive sensory nerves via a CB1 and CB2 cannabinoid receptor-independent mechanism. *J Neurosci*. 2002;22(11):4720−4727.

65. Engel MA, Izydorczyk I, Mueller-Tribbensee SM, Becker C, Neurath MF, Reeh PW. Inhibitory CB1 and activating/desensitizing TRPV1-mediated cannabinoid actions on CGRP release in rodent skin. *Neuropeptides*. 2011;45(3):229−237.

66. Thomas BF, Compton DR, Martin BR. Characterization of the lipophilicity of natural and synthetic analogs of delta 9-tetrahydrocannabinol and its relationship to pharmacological potency. *J Pharmacol Exp Ther*. 1990;255(2):624−630.

67. Gertsch J, Leonti M, Raduner S, et al. Beta-caryophyllene is a dietary cannabinoid. *Proc Natl Acad Sci U S A*. 2008;105(26):9099−9104.

68. Tiwari M, Kakkar P. Plant derived antioxidants - Geraniol and camphene protect rat alveolar macrophages against t-BHP induced oxidative stress. *Toxicol In Vitro*. 2009;23 (2):295−301.

69. Crowell PL. Prevention and therapy of cancer by dietary monoterpenes. *J Nutr*. 1999;129 (3):775S−778S.

70. Shoff SM, Grummer M, Yatvin MB, Elson CE. Concentration-dependent increase of murine P388 and B16 population doubling time by the acyclic monoterpene geraniol. *Cancer Res*. 1991;51(1):37−42.

71. d'Alessio PA, Ostan R, Bisson JF, Schulzke JD, Ursini MV, Bene MC. Oral administration of d-limonene controls inflammation in rat colitis and displays anti-inflammatory properties as diet supplementation in humans. *Life Sci*. 2013;92(24−26):1151−1156.

72. Barrett ML, Gordon D, Evans FJ. Isolation from *Cannabis sativa* L. of cannflavin—a novel inhibitor of prostaglandin production. *Biochem Pharmacol*. 1985;34(11):2019−2024.

73. Vasanthi HR, ShriShriMal N, Das DK. Phytochemicals from plants to combat cardiovascular disease. *Curr Med Chem*. 2012;19(14):2242−2251.

74. Nieman DC, Laupheimer MW, Ranchordas MK, Burke LM, Stear SJ, Castell LM. A-Z of nutritional supplements: dietary supplements, sports nutrition foods and ergogenic aids for health and performance—part 33. *Br J Sports Med*. 2012;46(8):618−620.

75. Cimino S, Sortino G, Favilla V, et al. Polyphenols: key issues involved in chemoprevention of prostate cancer. *Oxid Med Cell Longev*. 2012;2012:632959

76. Lübbert M, Kyereme J, Schöbel N, Beltrán L, Wetzel CH, Hatt H. Transient receptor potential channels encode volatile chemicals sensed by rat trigeminal ganglion neurons. *PLoS One*. 2013;8(10):e77998.

77. Ohkawara S, Tanaka-Kagawa T, Furukawa Y, Nishimura T, Jinno H. Activation of the human transient receptor potential vanilloid subtype 1 by essential oils. *Biol Pharm Bull*. 2010;33(8):1434−1437.

78. Stotz SC, Vriens J, Martyn D, Clardy J, Clapham DE. Citral sensing by transient [corrected] receptor potential channels in dorsal root ganglion neurons. *PLoS One*. 2008;3(5):e2082.

79. Schlosburg JE, O'Neal ST, Conrad DH, Lichtman AH. CB1 receptors mediate rimonabant-induced pruritic responses in mice: investigation of locus of action. *Psychopharmacology (Berl)*. 2011;216(3):323−331.

80. Schlosburg JE, Boger DL, Cravatt BF, Lichtman AH. Endocannabinoid modulation of scratching response in an acute allergenic model: a new prospective neural therapeutic target for pruritus. *J Pharmacol Exp Ther*. 2009;329(1):314−323.

CHAPTER *3*

Medical Cannabis Formulations

CANNABIS INFLORESCENCE AND HASHISH

Dried cannabis inflorescence (the complete flower head) is one of the most commonly encountered formulations for delivery of medicinal cannabinoids. In the Netherlands, for example, medicinal marijuana is grown by Bedrocan B.V. under contract with the Dutch Ministry of Health and made available as herbal inflorescence in Dutch pharmacies on prescription.[1] Bedrocan grows six varieties, and each is assayed and labeled with a standardized level of three cannabinoids: Δ^9-tetrahydrocannabinol (Δ^9-THC), cannabidiol (CBD), and cannabinol (CBN). The Ministry also exports cannabis for medicinal use to Italy, Germany, Finland, Canada, and the Czech Republic, and to approved researchers around the world. In Canada, there are 26 authorized licensed producers of dried cannabis and a more limited number of authorized and licensed producers of fresh cannabis and cannabis oil (Fig. 3.1). The flowering

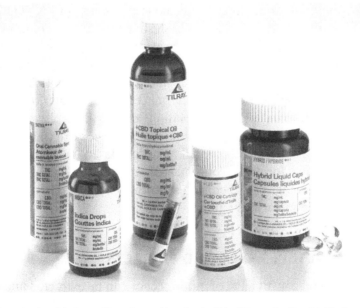

Figure 3.1 Example of medicinal cannabis products cultivated and sold by Tilray under Canada's Marihuana for Medical Purposes Regulations.

The Analytical Chemistry of Cannabis. DOI: http://dx.doi.org/10.1016/B978-0-12-804646-3.00003-5

tops or buds of a wide range of varieties are also available through dispensaries and medicinal programs in the United States and elsewhere and are sold under a surprisingly large number of popularized names such as "Skunk," "Kush," "Diesel," and "Haze." In contrast to the situation in the Netherlands, the differences in chemical content between these products are not often readily discernable from the label, or in some instances even through rigorous quantitative analysis, and the pharmacological and organoleptic effects can be unpredictable. The breadth of product lines in popular markets appears to be more a matter of marketing than of differential therapeutic utility.

Regardless of strain, when cannabis inflorescence is procured in bulk, it is often cut or further processed into finer material that can be rolled into "cigarettes" or otherwise used for combustion and inhalation administration. The cigarette form has historically been quite popular in the illicit market and is one of the most widely distributed dosage formulations in the US National Institute on Drug Abuse Drug Supply Program (NIDA DSP). Inflorescence can also be frozen and agitated to separate the phytocannabinoid-rich trichome resin head from the trichome stalk. The resin heads are then compressed into hashish, which is a physically refined form of cannabis that has been significantly enriched in phytocannabinoid content. Compared to plant material containing total concentrations of Δ^9-THCA and Δ^9-THC of up to 25% (w/w), hashish preparations commonly have concentrations of 65% (w/w). Alternative processing techniques include agitation of plant material in ice water followed by filtration separation of the resin heads through a sieve. The concentrated preparation of trichomes can be dried for pharmaceutical formulation and use or for further separation and isolation of purified phytocannabinoid constituents.[2] The freezing method can produce materials with higher Δ^9-THC content than traditional agitation-based methods. Like herbal cannabis, solid forms of concentrated trichomes vary considerably in shape, size, color, and chemical constituents, and are marketed under a variety of popularized names.

TEAS, TINCTURES, OILS, AND EXTRACTS

The chemical constituents of inflorescence can be extracted and purified further with various solvents. A simple method is the use of boiling water to make a tea (infusion) for oral administration. In a systematic laboratory study, Hazekamp et al. took a chemotype legally grown in the Netherlands with a Δ^9-THCA content of 191 mg/g (19.1%) and a Δ^9-THC content of 6 mg/g (0.6%) of dry weight plant material and prepared a standard tea using the procedures recommended by the Netherlands Office of Medicinal Cannabis. One gram of cannabis was added to 1 L of boiling water and simmered for 15 min. The remaining solids were removed with a common tea-sieve. Samples of the tea were lyophilized to complete dryness, reconstituted in ethanol, and analyzed using a validated HPLC method. Although the amount of Δ^9-THC in 1 g of inflorescence amounted to nearly 200 mg (the sum of Δ^9-THC and its precursor Δ^9-THCA), the whole 1 L volume of standard tea contained only a fraction of this amount (about 43 mg of Δ^9-THCA and 10 mg of Δ^9-THC) in the water phase. The relatively low concentration in the liquid was attributed to the limited solubility and saturation of the water phase with these lipophilic chemicals, as analysis of a precipitate remaining after the removal of solid plant material revealed that relatively complete extraction of the phytocannabinoids had actually occurred. Incomplete conversion of Δ^9-THCA into Δ^9-THC was also observed in the tea, as it was in experiments performed with pure standards.[3] Boiling for longer periods more than doubled the concentration of Δ^9-THC in the liquid, but did not result in complete conversion from Δ^9-THCA to Δ^9-THC after 30 min, or significantly increase the concentration of Δ^9-THCA in the liquid extract, again suggesting that thermal conversion rates and limited solubility under these conditions are the predominate factors affecting the final concentrations of phytocannabinoids in tea. It has been noted that the cannabinoid content in tea reported in the studies of Hazekamp et al. is unlike that found in the plant or the chemical exposures that occur with other dosage formulations, which is likely to affect bioavailability and pharmacological actions.[4] To compensate for the low solubility of Δ^9-THC in water, users of tea often add a small amount of vegetable oil or butter and increase boiling time to several hours.

Ethanol and other organic solvents are often more effective extraction agents than water and can be used to prepare fairly concentrated tinctures or extracts. Ethanolic tinctures have a long history of use through buccal or sublingual administration and ingestion. Extracts and tinctures appeared early in the United States Pharmacopoeia and other pharmaceutical references and were manufactured and marketed by pharmaceutical companies such as Merck in Germany, Burroughs-

Wellcome in Britain, and Eli Lilly in the United States.[5] In making tinctures, the plant material or the final product may be heated or otherwise treated to facilitate transformation of the biosynthetic acid forms into their neutral forms. Hot ethanol extraction has been used in the NIDA DSP to remove essentially all of the Δ^9-THCA, Δ^9-THC, and other cannabinoids from cannabis and produce both an extracted plant material that can be used as a placebo for research and a crude phytocannabinoid tincture that can be used as a drug product or in dosage formulations. To ensure complete extraction, the temperature of the ethanol is typically increased to approximately 60°C. The ethanol is recirculated for $\sim$24 h. After 24 h, the ethanol is replaced with fresh ethanol, and the process is repeated daily for at least 3 days. This approach can dramatically increase the total extraction and yield. However, ethanol is not a very selective extraction solvent, and many other chemical constituents are also removed. As a result, the placebo plant material burns more rapidly, and is often noted to differ considerably in its organoleptic properties from the nonextracted material. Nevertheless, the ethanol can be evaporated to produce a crude cannabis extract with a suitably high concentration of Δ^9-THCA and Δ^9-THC.

Chloroform, hexane, and other organic solvents that are heated and refluxed over time can be used to produce more refined extracts than can ethanol. However, many are toxic and must be carefully removed before product use. Compressed gases are readily available to produce highly concentrated preparations. For example, butane extracts are popular in dispensaries and on the illicit market. But doing the extraction improperly can produce explosions and fires leading to severe burns, other injuries, and death. A safer approach is supercritical fluid extraction with compressed liquid CO_2. Cannabinoids can be isolated from other plant constituents by immersion in liquid CO_2 or the use of commercially available supercritical fluid extraction units, providing a relatively inexpensive and facile approach to purification of large amounts of relatively pure final products. However, while CO_2 is not flammable, there are risks from asphyxiation and high-pressure failures. In addition, if Δ^9-THC or CBD is the desired end product to be obtained by the extraction, further heating is typically required to ensure that all of the acid forms are decarboxylated to their neutral analogs. Because of the relative ease and efficiency of extraction, liquid CO_2 is increasingly employed for purification of phytocannabinoids and has rekindled interest in supercritical fluid extraction in the isolation of chemical constituents from other natural products. For example, the pharmaceutical formulation Sativex is manufactured in the United Kingdom by incorporating purified botanically derived drug substances containing Δ^9-THC and CBD (initially isolated from cannabis inflorescence with liquid CO_2) in an accurately measured ratio into a final drug product containing ethanol, propylene glycol, and peppermint oil as excipients.[6]

CONSUMABLES

Edible or ingested cannabis products are popular. They are produced and marketed in a wide variety of forms, including solids that are eaten, liquids that are swallowed, and mints, candies, lollipops, and other products to be held in the mouth for buccal absorption (Fig. 3.2). As with other formulations, there is potential for variance in chemical composition, strength, and homogeneity based on the raw plant material variety and the method of manufacture. For example, when cannabis inflorescence or extract is used in the preparation of food products baked at high temperature, the cannabinoids will be decarboxylated to their neutral forms, a desirable quality for pharmacological effects. This is not the case with popsicles, soft drinks, and other products not subjected to high heat, so the manufacturer may use plant material that has been aged or heated. It should be emphasized that the ingredients used in the preparation of consumables are the equivalent of excipients in pharmaceutical preparations, and that both active ingredients and excipients should be carefully selected, tested for compatibility and storage stability, and analytically monitored and assessed during manufacture and release of the final product. The variety of consumable dosage formulations made available to patients through cannabis dispensaries might be better suited for consistency in chemical constituents and pharmacological effect if they were more limited in nature and number and subjected to well-described manufacturing specifications and quality control.

Product images courtesy of the Werc Shop

Figure 3.2 Example of edible medicinal cannabis products purchased in dispensaries in Seattle, Washington, USA.

Figure 3.2 (Continued).

FORMULATIONS FOR PARENTERAL ADMINISTRATION

Δ^9-THC and other phytocannabinoids have been formulated for transmucosal, transdermal, topical (eg, ophthalmic), rectal, and intravenous (i.v.) routes.[7] Sublingual and intranasal transmucosal continue to be popular for their ease of administration and rapid onset of action. One of the best-characterized oromucosal preparations is Sativex, a metered dose spray manufactured by GW Pharmaceuticals and approved for the spasticity of multiple sclerosis. It is also in development for cancer pain. According to the patient information leaflet, each 0.1 mL spray contains 3.8 to 4.4 mg and 3.5 to 4.2 mg of two extracts (as soft extracts) from *Cannabis sativa L.* leaf and flower, corresponding to 2.7 mg Δ^9-THC and 2.5 mg CBD, dissolved in the excipients ethanol, propylene glycol, and peppermint oil. This product is available in 15 countries and has been approved in 12 others but has not been approved by the US Food and Drug Administration (FDA) for sale in the United States. However, both Sativex and Epidiolex 1, an investigational CBD-containing cannabis-derived oromucosal product being developed by GW Pharmaceuticals for epilepsy, received Fast Track Designation by the FDA for specific therapeutic indications. The tests with Sativex in cancer patients have not produced results superior to placebo, but the FDA might still approve it for multiple sclerosis. Mouth sprays, drops, and gums containing Δ^9-THC and other cannabinoids are becoming increasingly common in cannabis dispensaries, where they are being offered in a wide variety of concentrations and flavors, including cinnamon, peppermint, and fruits.

Intranasal transmucosal formulations have been tested in laboratory animals and deemed viable. For example, one containing glatiramer acetate and CBD in soy phospholipid (Phospholipon 90 G) containing 94% phosphatidylcholine, lysophosphatidylcholine, and vitamin E has been reported effective in experimental autoimmune encephalomyelitis, an animal model of multiple sclerosis. The drug combination reduced the clinical signs better than each alone and augmented neuron regeneration in the hippocampus.[8] Several other intranasal formulations, including mucoadhesive gels and permeability enhancers, have been evaluated.[9–11] There appears to be continuing interest and merit in intranasal products, which are noticeably absent in most medicinal cannabis dispensaries.

A variety of transdermal formulations have been shown to produce pharmacological effects locally and systemically.[12,13] For example, an appliance (7.25 cm^2) containing 8 mg of Δ^8-THC in 0.5 mL 1:1:1 (v/v/v) propylene glycol:water:ethanol in a drug reservoir sandwiched between a drug-impermeable backing laminate (MEDIFLEX 1502) and a rate-controlling EVA membrane (CoTran 9715) with ARcare7396 adhesive showed sustained systemic transdermal drug delivery.[14] The skin permeability enhancer 2-(2-ethoxyethoxy)ethanol in similar CBD-containing formulations increased plasma steady-state levels almost fourfold.[10] Sustained transdermal delivery systems offer considerable advantages over other formulations and continue to be of therapeutic interest, but are not often found in medical cannabis dispensaries. Instead, salves and creams containing cannabinoids suspended in coconut oil, beeswax, shea butter, and other excipients intended for local application are common.

Δ^9-THC in light mineral oil or olive oil has been tested for topical ophthalmic use in glaucoma. Though effective in laboratory animals,[15,16] 0.05% and 0.1% in light mineral oil were no more effective than vehicle control in six human subjects with primary open-angle glaucoma in a randomized, balanced, double-masked protocol.[16] Similar lack of efficacy was noted by other investigators for light mineral oil formulations,[17] which has led to the development and testing of alternative topical formulations for ocular application using submicron emulsions,[18,19] cyclodextrins,[20,21] ion-pairing reagents,[22] hydroxypropylmethylcellulose,[21] polyvinyl alcohol, polyvinylpyrrolidone, corn oil, hyaluronic acid,[23] and alternative[24] or prodrug forms of phytocannabinoids[25] in order to improve drug solubility, stability, transcorneal permeability, and efficacy. Some improvements have been demonstrated in drug delivery and efficacy; however, the availability of superior therapeutic agents has limited the development and use of phytocannabinoid-containing topical formulations for glaucoma and other ocular diseases, and ophthalmic products are not generally offered by cannabis dispensaries.

Alternative formulations include rectal suppositories as well as spray freeze dried formulations to be inhaled as with asthma inhalers. For example, the bioavailability of Δ^9-THC from suppositories containing prodrug hemisuccinate, N-formyl alaninate, N-methyl carbamate, and methoxy acetate esters in both lipophilic (Witepsol H15) and hydrophilic (polyethylene glycol) bases was tested in laboratory animals.[26] Among these formulations, the use of the hemisuccinate ester of Δ^9-THC and Witepsol was found to provide the most sustained and consistent release and delivery and was well tolerated in limited clinical studies.[27,28] Regardless of their potential pharmaceutical advantages, at present these formulations have more interest in research than in the commercial market. Nevertheless, the evolution and popularity of novel cannabis-derived formulations is increasing at a remarkable pace to include lip balms, shampoos, massage oils, toothpastes, sunscreens, lubricants, and products for pets.

Finally, the i.v. route of cannabinoid administration is generally reserved for nonclinical laboratory settings; hence, many i.v. formulations have been prepared in a manner that is not consistent with human use. Human use-compatible formulations can be prepared using solvents such as ethanol and propylene glycol, or cyclodextrin and other excipients that serve as carriers for the lipophilic cannabinoids and facilitate their solubility or suspension in sterile saline. The NIDA Drug Supply provides resources for the formulation and manufacture of sterile injectable formulations for use in nonclinical and clinical research. However, because of the inherent risks of infection associated with the intravenous, intramuscular, and subcutaneous routes, these formulations are infrequently requested and seldom prepared for human use. Ethanolic solutions of phytocannabinoids are requested and provided for formulation by the investigator for laboratory animal experiments or the occasional human use study. When these ampules of cannabinoids in ethanol are provided for such studies, the ethanol is typically evaporated to dryness and a suitable sterile formulation is prepared by the researcher in accordance with institutional, state, and federal requirements and procedures for experimental protocol review, approval, and use (eg, Institutional Animal Care and Use Committee, Investigational New Drug Application, Institutional Review Board), as is the case with any new dosage formulation that is tested in nonclinical or clinical studies. Unfortunately, similar review and approval of novel dose formulations and their intended use are not generally required in cannabis dispensaries, and good manufacturing and quality control standards are not uniformly applied and enforced across the wide variety of drug products being offered.

SMOKING AND VAPORIZING

Regardless of the cannabis raw material or formulation selected for use, the route of administration, unit dose, and dose frequency are often left up to the patient or recreational user to determine what is best suited to their needs. Two commonly employed routes for herbal inflorescence are inhalation of smoke after combustion in pipes or cigarettes, and inhalation of aerosols and gas/vapor streams after heating in noncombusting appliances. However, it is important to emphasize that the chemical constituents in cannabis are not representative of the chemicals present in smoke. This is because the heating and incineration of the plant material produces a variety of thermolysis and combustion products in the inhaled smoke. Of particular importance to a medicinal cannabis user seeking the pharmacological effects of Δ^9-THC is the thermolytic conversion of Δ^9-THCA into Δ^9-THC, which occurs rapidly such that about 70% of the Δ^9-THCA is converted into Δ^9-THC prior to inhalation of the smoke into the mouth and lungs (approximately 30% of the total THC equivalents are lost to oxidative degradation during combustion and heating). Studies have suggested that the thermal conversion of Δ^9-THCA to Δ^9-THC serves to facilitate the delivery of Δ^9-THC, and that more thermal decomposition and lower smoke yields of Δ^9-THC may occur if the Δ^9-THCA in the plant material is converted to Δ^9-THC prior to smoking.[29]

Once inhaled, the decarboxylated phytocannabinoids are absorbed into the bloodstream, producing rapid increases in their concentration in circulation to peripheral tissues. As the flow of blood from the lungs avoids first-pass metabolism occurring in the liver, plasma concentration profiles for phytocannabinoids such as Δ^9-THC are similar to those achieved with i.v. administration.[30] However, the absorption and bioavailability following the smoking route has been found to be quite variable (between 2% and 56%), due in part to intra- and intersubject variability in smoking dynamics, loss of chemical constituents to side stream smoke, and other factors contributing to uncertainty in dose delivery.[31–33] Despite this variability, the increase in plasma concentration occurs relatively rapidly after inhalation, and the circulating drug then gains rapid access to the brain, providing a reasonably timely feedback between dosing parameters and effect that allows dose titration and helps prevent overdosing.

The combustion and heating of cannabis promotes decarboxylation of Δ^9-THCA and CBDA to Δ^9-THC and CBD, which are commonly considered the desired, pharmacologically active end products for inhalation. Unfortunately, other chemical products of combustion that are inhaled include the "tar" and hazardous substances that are formed during the combustion of all plant material, carbon monoxide (a predominant component), and polycyclic aromatic hydrocarbons. As compared to filtered tobacco cigarette smokers, marijuana smokers experience a fivefold greater increment in their blood carboxyhemoglobin levels, and an approximately threefold increase in inhaled tar; one-third more of the inhaled tar is retained in the respiratory tract. These increased exposures are associated with differences noted experimentally in the dynamics of smoking cannabis and tobacco, among them an approximately two-thirds larger puff volume, a one-third increased depth of inhalation, and a fourfold longer duration of inhalation hold time with cannabis than with tobacco.[34] Another recent study on the health impact of cannabis smoke exposure demonstrated that short exposure to marijuana secondhand smoke causes acute vascular endothelial dysfunction (impaired arterial flow-mediated dilation) similar to that caused by tobacco secondhand smoke. The investigators also observed similar endothelial effects with exposure to secondhand smoke from placebo marijuana lacking Δ^9-THC, demonstrating that impairment was due to smoke constituents other than Δ^9-THC.[35] These untoward exposures may be diminished by the use of filtration devices such as water pipes, or through the use of vaporization instead of combustion. As a result of the increased appreciation of the inherent risks associated with combustion of cannabis and inhalation of smoke at present, bulk cannabis inflorescence is also provided through the NIDA DSP for use in other formulations and non-combusted routes of administration such as vaporization.

Compared to inhalation of smoke from combusted cannabis, vaporization of cannabis inflorescence or cannabis-derived formulations can significantly reduce the harmful exposures. The heating of the material still serves to promote the decarboxylation of the phytocannabinoid acids and the delivery of the constituents in the aerosol and gas/vapor phase, but heating can also produce oxidative degradation such as the conversion of Δ^9-THC to CBN, isomerization of Δ^9-THC to Δ^8-THC, and inhalation exposure to a variety of unintended and undesirable chemicals. In addition, as with smoking, the dose of phytocannabinoids and other chemical constituents that is delivered by vaporization and inhalation can vary considerably depending upon the formulation used and the design and configuration of the vaporizer and its heating element. For example, when herbal inflorescence is heated, the concentration of the phytocannabinoids in the plant material is one of the primary determinants of the amount of these constituents that is delivered to the lungs. The volume and depth of inhalation and the retention of the inhaled aerosol and gas/vapors in the lung for an extended period of time can also influence the delivery and extraction of active ingredients into the lungs and bloodstream. This is also the case when extracts of phytocannabinoids are heated and inhaled with e-cigarettes or pipes. A matter of particular concern with vaporization is that in some cases the extracts available for consumers can be over 80% pure Δ^9-THC, which can result in administration of a significantly greater dose than could be achieved with herbal inflorescence or most forms of hashish. The exposure of naïve individuals to such high inhaled concentrations of active cannabinoids can bypass the ability of the user to titrate the dose, which can lead to severe intoxication and discomfort.

BIOAVAILABILITY FOR ENTERAL AND TRANSMUCOSAL ADMINISTRATION

The bioavailability of Δ^9-THC, CBD, and other phytocannabinoids is generally low and variable in humans after ingestion.[31] Cannabis tea and other products that are ingested undergo first-pass metabolism by the liver as they are absorbed from the gastrointestinal tract, whereas tinctures or liquid formulations used sublingually are absorbed into the bloodstream without first-pass metabolism. The acidic milieu of the stomach can cause chemical transformation of phytocannabinoids, including conversion of CBD into Δ^9-THC and isomerization of Δ^9-THC into Δ^8-THC. As noted with Marinol (dronabinol) and other relatively simple formulations of Δ^9-THC, markedly different phytocannabinoid absorption and pharmacological properties may be observed between and within ingested products due to differences in product composition (vehicles or ingredients) and individual variability (body mass, diet and stomach contents, gastric emptying times, etc.). For example, when gelatin capsules containing liquid formulations were formulated with similar phytocannabinoid content and orally administered to humans, the speed and degree of absorption was greatly influenced by the liquid vehicle, and even when the same vehicle was used, absorption of the drug varied significantly among individuals.[36] Thus, even with a consistently produced formulation of synthetically derived Δ^9-THC in sesame oil (Marinol), absorption of the Δ^9-THC can vary considerably due to differences in individuals and their diets. More importantly, the time lag between dose administration and effect compared to inhalation routes can be considerable. This can lead to inadvertent overdosing, as individuals may take additional doses before the absorption and pharmacological effects from the initial exposure have reached their maximum level. Improved formulations of orally ingested Δ^9-THC are still of therapeutic interest, and are currently being developed and tested in clinical trials. For example, Namisol is a novel tablet formulation of Δ^9-THC and Alitra (Echo Pharmaceuticals b.v., Nijmegen, the Netherlands), an emulsifying drug delivery technology, which has been studied in human trials using sublingual and oral routes.[4] Despite the limited sample size, the data suggest that Δ^9-THC from Namisol might have a faster absorption rate and reach a less variable maximal concentration, and therefore be more favorable than currently registered oral dronabinol formulations and cannabis-based medicines with respect to its pharmacokinetic and pharmacodynamic properties.

Whether intranasal or oral sprays or drops are used, the transmucosal delivery of cannabis-derived therapeutics has the advantage of bypassing metabolism in the gut and liver. In addition, it avoids the confounding effects of stomach contents, gastrointestinal emptying times, and the chemically reactive acidic conditions encountered in the stomach. Therefore, the delivery of cannabinoids is typically more predictable when a transmucosal spray is used, particularly with a metered dose delivery system. In addition, the absorption and delivery of cannabinoids from these formulations is relatively faster than from ingested formulations like Marinol, but slower than that occurring after inhalation of smoked or vaporized and inhaled products. This tends to diminish the risks that individuals will dose repeatedly prior to the time of maximum effect from the initial dose, which is a continuing problem for cannabis edibles. With Sativex, the package insert is very clear in instructing users to carefully titrate the dose, particularly as they begin treatment. Specifically, the package insert states that plasma starts to contain measurable levels of cannabinoids and their metabolites after 30 min, and advises that treatment should be started at a maximum rate of one spray every 4 h on the first day, up to a maximum of four sprays on the first day. This dose titration strategy is consistent with the pharmacokinetic information provided in the insert, which indicates that the T_{max}, or the time after administration of a drug when the maximum plasma concentration is reached, is between 2 and 4 h. However, the insert also warns that individual subject plasma concentration data and pharmacokinetic parameters show a high degree of intersubject variability, and emphasizes that the product can produce an overdose that requires immediate medical attention. It is possible to limit the number of sprays that are delivered from a metered dosing device, which would further serve to decrease the likelihood of misuse or accidental overdose. However, the currently available spray formulations of cannabinoids do not generally employ this approach, as it appears the risks associated with misuse, overdose, or abuse and dependence with the existing device do not warrant the added expense involved in producing an alternative and safer drug delivery device.

BIOAVAILABILITY FOR OTHER PARENTERAL ROUTES OF ADMINISTRATION

Transdermal delivery of phytocannabinoids using a well-designed patch can provide relatively constant plasma levels for an extended period of time. For example, a study of a Δ^9-THC-containing transdermal patch in guinea pigs demonstrated that a mean steady-state plasma level of 4.4 ng/mL could be reached within 1.4 h and maintained for at least 48 h, and that significant amounts of metabolites accumulated in the plasma after extended topical application.[14] After removal of the patch, sustained delivery has been noted to continue for up to 24 h, which suggests that there is a considerable depot of active drug in the skin and subcutaneous tissue that slowly decreases over time. This depot effect may be due to the lipophilic nature of phytocannabinoids, which increases retention in lipid membranes of the skin and in fat. It has also been shown that phytocannabinoids and the endocannabinoid system are involved in skin sensation and cutaneous cell growth and differentiation,[37] that lipophilic cannabinoids are sequestered in sebaceous glands,[38] and that CBD can modulate sebaceous gland function and inflammation and act as a highly effective sebostatic agent.[39] Thus phytocannabinoid formulations for topical application may prove to be of therapeutic value for indications such as acne, seborrhea, allergic dermatitis, pruritus (itch), pain, psoriasis, hair growth disorders, systemic sclerosis, and cancer.[40]

Rectal suppository formulations containing Δ^9-THC-hemisuccinate ester have the advantage of bypassing metabolism by the gut and liver, and have been shown to produce increased delivery and bioavailability of cannabinoids in laboratory animals and superior therapeutic effects over oral ingestion in limited clinical studies. In general, plasma levels increase within 15 to 30 min and can produce effects that are sustained over 4 to 8 h.[7,27,28] Though a variety of rectal suppositories containing cannabinoids are manufactured and marketed in cannabis dispensaries and on the internet, this route appears not to be widely employed at the present time.

Intravenous and other parenteral routes can provide the most consistent delivery of cannabinoids, and can often produce the lowest inter- and intraindividual variability in drug plasma concentrations.[30−32] After a bolus i.v. injection, concentrations of cannabinoids in circulating blood rapidly rise and then fall due to distribution into tissues, with metabolism and elimination of parent compound progressing at slower rates.[30−33] Other injection routes, such as intramuscular and subcutaneous, provide for a slower, more variable increase in plasma levels of cannabinoids but can achieve a more sustained delivery of cannabinoids and their metabolites than a bolus i.v. injection. There are also ongoing efforts aimed at designing sustained long-release depot formulations of cannabinoids that can be injected subcutaneously or intramuscularly.[41−43] For example, the development of Δ^9-THC biodegradable microspheres as an alternative delivery system for cannabinoid parenteral administration has been reported. In this study, tetrahydrocannabinol was encapsulated into biodegradable microspheres by the oil-in-water (o/w) emulsion solvent evaporation method, and several formulations were prepared using different drug:polymer ratios and antioxidant excipients. The in vitro drug release studies showed that the encapsulated drug was released over a 2-week period and demonstrated efficacy in inhibiting the growth of a cancer cell line.[41] Similarly, sustained release formulations of CBD- and Δ^9-THC-loaded poly-ε-caprolactone microparticles have produced long-term cannabinoid administration and enhanced apoptosis and decreased cell proliferation and angiogenesis in a murine xenograft model of glioma.[42] The merits of parenteral routes of administration are apparent, but the risks associated with infection, overdose, or misinjection and injury generally relegates this mode to laboratory animals or medical professionals.

REFERENCES

1. Bedrocan. <http://www.bedrocan.nl/english/home.html>; 2015 Accessed 28.09.15.

2. Potter DJ. Cannabis horticulture. In: Pertwee RG, ed. *Handbook of Cannabis*. New York, NY: Oxford University Press; 2014:65–88.

3. Hazekamp A, Bastola K, Rashidi H, Bender J, Verpoorte R. Cannabis tea revisited: a systematic evaluation of the cannabinoid composition of cannabis tea. *J Ethnopharmacol.* 2007;113(1):85–90.

4. Klumpers LE, Beumer TL, van Hasselt JG, et al. Novel Delta(9)-tetrahydrocannabinol formulation Namisol® has beneficial pharmacokinetics and promising pharmacodynamic effects. *Br J Clin Pharmacol.* 2012;74(1):42–53.

5. Zuardi AW. History of cannabis as a medicine: a review. *Rev Bras Psiquiatr.* 2006; 28:153–157.

6. Potter DJ. A review of the cultivation and processing of cannabis (*Cannabis sativa* L.) for production of prescription medicines in the UK. *Drug Test Anal.* 2014;6(1–2):31–38.

7. Grotenhermen F. Cannabinoids for therapeutic use. *Am J Drug Deliv.* 2004;2(4):229–240.

8. Duchi S, Ovadia H, Touitou E. Nasal administration of drugs as a new non-invasive strategy for efficient treatment of multiple sclerosis. *J Neuroimmunol.* 2013;258(1–2):32–40.

9. Al-Ghananeem AM, Malkawi AH, Crooks PA. Bioavailability of Delta(9)-tetrahydrocannabinol following intranasal administration of a mucoadhesive gel spray delivery system in conscious rabbits. *Drug Dev Ind Pharm.* 2011;37(3):329–334.

10. Paudel KS, Hammell DC, Agu RU, Valiveti S, Stinchcomb AL. Cannabidiol bioavailability after nasal and transdermal application: effect of permeation enhancers. *Drug Dev Ind Pharm.* 2010;36(9):1088–1097.

11. Valiveti S, Agu RU, Hammell DC, et al. Intranasal absorption of Delta(9)-tetrahydrocannabinol and WIN55,212-2 mesylate in rats. *Eur J Pharm Biopharm.* 2007;65(2):247–252.

12. Cichewicz DL, Welch SP, Smith FL. Enhancement of transdermal fentanyl and buprenorphine antinociception by transdermal delta9-tetrahydrocannabinol. *Eur J Pharmacol.* 2005;525(1–3):74–82.

13. Liput DJ, Hammell DC, Stinchcomb AL, Nixon K. Transdermal delivery of cannabidiol attenuates binge alcohol-induced neurodegeneration in a rodent model of an alcohol use disorder. *Pharmacol Biochem Behav.* 2013;111:120–127.

14. Valiveti S, Hammell DC, Earles DC, Stinchcomb AL. In vitro/in vivo correlation studies for transdermal delta 8-THC development. *J Pharm Sci.* 2004;93(5):1154–1164.

15. Fischer KM, Ward DA, Hendrix DV. Effects of a topically applied 2% delta-9-tetrahydrocannabinol ophthalmic solution on intraocular pressure and aqueous humor flow rate in clinically normal dogs. *Am J Vet Res.* 2013;74(2):275–280.

16. Merritt JC, Perry DD, Russell DN, Jones BF. Topical delta 9-tetrahydrocannabinol and aqueous dynamics in glaucoma. *J Clin Pharmacol.* 1981;21(8–9 suppl):467S–471S.

17. Green K, Roth M. Ocular effects of topical administration of delta 9-tetrahydrocannabinol in man. *Arch Ophthalmol (Chicago, Ill.: 1960)*. 1982;100(2):265–267.

18. Naveh N, Weissman C, Muchtar S, Benita S, Mechoulam R. A submicron emulsion of HU-211, a synthetic cannabinoid, reduces intraocular pressure in rabbits. *Graefes Arch Clin Exp Ophthalmol*. 2000;238(4):334–338.

19. Merritt JC, Olsen JL, Armstrong JR, McKinnon SM. Topical delta 9-tetrahydrocannabinol in hypertensive glaucomas. *J Pharm Pharmacol*. 1981;33(1):40–41.

20. Hippalgaonkar K, Gul W, ElSohly MA, Repka MA, Majumdar S. Enhanced solubility, stability, and transcorneal permeability of delta-8-tetrahydrocannabinol in the presence of cyclodextrins. *AAPS PharmSciTech*. 2011;12(2):723–731.

21. Green KE, Kearse CE. Ocular penetration of topical Delta9-tetrahydrocannabinol from rabbit corneal or cul-de-sac application site. *Curr Eye Res*. 2000;21(1):566–570.

22. Hingorani T, Gul W, Elsohly M, Repka MA, Majumdar S. Effect of ion pairing on in vitro transcorneal permeability of a Delta(9)-tetrahydrocannabinol prodrug: potential in glaucoma therapy. *J Pharm Sci*. 2012;101(2):616–626.

23. Kearse EC, Green K. Effect of vehicle upon in vitro transcorneal permeability and intracorneal content of Delta9-tetrahydrocannabinol. *Curr Eye Res*. 2000;20(6):496–501.

24. Colasanti BK. A comparison of the ocular and central effects of delta 9-tetrahydrocannabinol and cannabigerol. *J Ocul Pharmacol*. 1990;6(4):259–269.

25. Hingorani T, Adelli GR, Punyamurthula N, et al. Ocular disposition of the hemiglutarate ester prodrug of (9)-tetrahydrocannabinol from various ophthalmic formulations. *Pharm Res*. 2013;30(8):2146–2156.

26. Elsohly MA, Little Jr. TL, Hikal A, Harland E, Stanford DF, Walker L. Rectal bioavailability of delta-9-tetrahydrocannabinol from various esters. *Pharmacol Biochem Behav*. 1991;40 (3):497–502.

27. Brenneisen R, Egli A, Elsohly MA, Henn V, Spiess Y. The effect of orally and rectally administered Δ 9-tetrahydrocannabinol on spasticity: a pilot study with 2 patients. *Int J Clin Pharmacol Ther*. 1996;34(10):446–452.

28. Mattes RD, Shaw LM, Edling-Owens J, Engelman K, Elsohly MA. Bypassing the first-pass effect for the therapeutic use of cannabinoids. *Pharmacol Biochem Behav*. 1993;44 (3):745–747.

29. Dussy FE, Hamberg C, Luginbühl M, Schwerzmann T, Briellmann TA. Isolation of Δ9-THCA-A from hemp and analytical aspects concerning the determination of Δ9-THC in cannabis products. *Forensic Sci Int*. 2005;149(1):3–10.

30. Ohlsson A, Lindgren JE, Wahlen A, Agurell S, Hollister LE, Gillespie HK. Plasma delta-9 tetrahydrocannabinol concentrations and clinical effects after oral and intravenous administration and smoking. *Clin Pharmacol Ther*. 1980;28(3):409–416.

31. Agurell S, Halldin M, Lindgren JE, et al. Pharmacokinetics and metabolism of delta 1-tetrahydrocannabinol and other cannabinoids with emphasis on man. *Pharmacol Rev*. 1986;38(1):21–43.

32. Huestis MA. Human cannabinoid pharmacokinetics. *Chem Biodivers*. 2007;4(8):1770–1804.

33. Ohlsson A, Lindgren JE, Wahlen A, Agurell S, Hollister LE, Gillespie HK. Single dose kinetics of deuterium labelled delta 1-tetrahydrocannabinol in heavy and light cannabis users. *Biomed Mass Spectrom*. 1982;9(1):6–10.

34. Wu T-C, Tashkin DP, Djahed B, Rose JE. Pulmonary hazards of smoking marijuana as compared with tobacco. *N Engl J Med*. 1988;318(6):347–351.

35. Wang X, Derakhshandeh R, Narayan S, et al. Abstract 19538: brief exposure to Marijuana secondhand smoke impairs vascular endothelial function. *Circulation*. 2014;130(suppl 2): A19538.

36. Perez-Reyes M, Lipton MA, Timmons MC, Wall ME, Brine DR, Davis KH. Pharmacology of orally administered Δ9-tetrahydrocannabinol. *Clin Pharmacol Ther*. 1973;14(1):48–55.

37. Caterina MJ. TRP channel cannabinoid receptors in skin sensation, homeostasis, and inflammation. *ACS Chem Neurosci*. 2014;5(11):1107–1116.

38. Lynn AB, Herkenham M. Localization of cannabinoid receptors and nonsaturable high-density cannabinoid binding sites in peripheral tissues of the rat: implications for receptor-mediated immune modulation by cannabinoids. *J Pharmacol Exp Ther*. 1994;268 (3):1612–1623.

39. Olah A, Toth BI, Borbiro I, et al. Cannabidiol exerts sebostatic and antiinflammatory effects on human sebocytes. *J Clin Invest*. 2014;124(9):3713–3724.

40. Biro T, Toth BI, Hasko G, Paus R, Pacher P. The endocannabinoid system of the skin in health and disease: novel perspectives and therapeutic opportunities. *Trends Pharmacol Sci*. 2009;30(8):411–420.

41. De La Ossa DHP, Gil-Alegre ME, Ligresti A, et al. Preparation and characterization of Δ9-tetrahydrocannabinol-loaded biodegradable polymeric microparticles and their antitumoral efficacy on cancer cell lines. *J Drug Target*. 2013;21(8):710–718.

42. Hernán Pérez de la Ossa D, Lorente M, Gil-Alegre ME, et al. Local delivery of cannabinoid-loaded microparticles inhibits tumor growth in a murine xenograft model of glioblastoma multiforme. *PLoS One*. 2013;8(1).

43. Martín-Banderas L, Muñoz-Rubio I, Prados J, et al. In vitro and in vivo evaluation of Δ9-tetrahidrocannabinol/PLGA nanoparticles for cancer chemotherapy. *Int J Pharm*. 2015;487 (1–2):205–212.

CHAPTER 4

Analytical Methods in Formulation Development and Manufacturing

With all cannabis and cannabis-derived formulations, there is the potential for considerable variance in chemical composition, strength, and homogeneity based on the selection of plant material seed stock and the methods used in their growth, harvesting, processing, formulation, and manufacture.[1–4] The chemotype or metabolome of a cannabis plant results from the complex integrated biosynthetic pathways for cannabinoids, terpenoids, flavonoids, and other constituents that are defined by genetics and heavily influenced by environmental factors. As the chemical constituents of cannabis can vary dramatically, accurate analytical information on the chemical content and strength of the natural product is required for its safe and effective use as a drug product or as a raw material for formulation. Information on the content of many different chemical constituents and potential contaminants may be required in order to determine suitability for use. In addition, the analytical characterization of chemical and physical factors, including moisture content, blend or batch homogeneity, and content uniformity, can play a particularly important role in formulations of cannabis inflorescence that are combusted or vaporized and inhaled (including hashish and hash oils, extracts, and concentrates), or manufactured into edible products, tablets, capsules, tinctures, salves, buccal or sublingual sprays, and other dosage forms. Furthermore, as cannabis is a natural product that can harbor many kinds of pests and microbes, including bacteria and fungi, careful analytical monitoring and quality control to exclude these and other potentially harmful contaminants from medicinal products is also a necessity. In some federally administered medicinal cannabis programs, for example, products are independently tested for appearance, cannabinoid and terpene profile, and moisture content, and for the absence of heavy metals, pesticides, bacteria, molds, and fungal toxins.[5]

The Analytical Chemistry of Cannabis. DOI: http://dx.doi.org/10.1016/B978-0-12-804646-3.00004-7

GENERAL CONSIDERATIONS IN SAMPLE PREPARATION FOR ANALYTICAL CHARACTERIZATION

A variety of sample preparation, separation, and detection methods have been shown to be useful in the characterization and standardization of cannabis inflorescence and cannabis-derived drug products.[6-18] But it should be emphasized that no single extraction solvent will extract all chemicals present, nor can a single means of separation and detection adequately isolate, detect, and define the complexity of the chemicals in the sample; these limitations hold true for all analytical procedures, including those based on mass spectrometry (MS) and nuclear magnetic resonance (NMR). Therefore, preparation and extraction must be carefully considered and performed under controlled conditions to ensure that reliable characterization of desired constituents occurs while precluding contamination of the sample.

Samples of cannabis inflorescence or dosage units should be taken based on sampling protocols that allow batch or blend homogeneity and content uniformity to be assessed as recommended by best practice and guidance (eg, see the World Health Organization's Quality Control Methods for Medicinal Plant Materials,[19] the US FDA Draft Guidance for Botanical Drug Development,[20] and the United States Pharmacopeia—National Formulary [USP-NF][21]). The sampling approach used should take into consideration batch size, suspected or past data on material homogeneity, intended use, requirements for amount of material needed to complete all analytical procedures and for retainer, and other factors, and the material should be frozen or processed immediately to avoid any biochemical change. After appropriate sampling has occurred and been documented, the samples are typically ground or otherwise processed individually into homogeneous particles for extraction and analysis. The apparatus used for grinding must then be carefully rinsed to recover any material adhering to the equipment and prevent contamination of subsequent samples. Plant material can also be directly homogenized in acetonitrile or other solvents using a Polytron or other homogenization device to process the bulk material into fine particle sizes while they are being mixed and extracted by the solvent. This approach decreases the loss of the resinous trichomes and their associated cannabinoid content that may occur with other approaches to grinding and homogenization. A wide range of methods for sample extraction are possible, each requiring a judicious selection of conditions such as temperature, pressure, and time. Internal standards should be used when available to help control for the recovery of certain chemicals for quantitative purposes, particularly as the solvent extraction processes can lead to varying degrees of decarboxylation and degradation of the phytocannabinoids.[22] As terpenes (and cannabinoids) are known to be chemically unstable during conditions encountered in many extraction techniques,[23] supercritical fluid extraction using liquid CO_2 or compressed gases offers several advantages, including effectiveness at low temperature and the reduction or elimination of thermal degradation and residual solvents before analysis.[9,13,24,25] Care is required in the use of compressed gases for extraction. For example, CO_2 can lead to asphyxiation and death if not handled properly. Butane has also become popular for the extraction of cannabis for consumption; however, the dangers involved in using this compressed gas have been well documented on YouTube and in the popular media.

DIRECT ANALYSIS OF CANNABIS INFLORESCENCE AND ITS EXTRACTS

Direct analysis of solids or crude extracts by $MS^{14,26}$ or $NMR^{14,27}$ can provide information in the absence of chromatographic resolution. NMR can be particularly useful for the direct analysis of cannabis extracts, and has several advantages over other analytical approaches for qualitative and quantitative characterization of complex mixtures.[12] For instance, it can give a detailed analysis of chemical composition very quickly with relatively simple sample preparation; it is a universal detector for molecules containing NMR-active nuclei; and the intensity of all proton signals is proportional to the molar concentration of the metabolite. With internal standardization, the concentration of chemical constituents can be determined accurately and with high reproducibility. ^1H-NMR has been used to demonstrate that the temperature and polarity of solvents used for the extraction of cannabis affect the total amount of Δ^9-THC in the extracts and its relative quantity with respect to Δ^9-THCA and other metabolites.[27] This study also showed that ^1H-NMR, without any evaporation or separation procedure, can distinguish between tinctures manufactured from different cannabis cultivars (Fig. 4.1).[27] In an extended study combining NMR and high-pressure liquid chromatography (HPLC) with diode array detection, a collection of chemical markers was identified in herbal drugs and

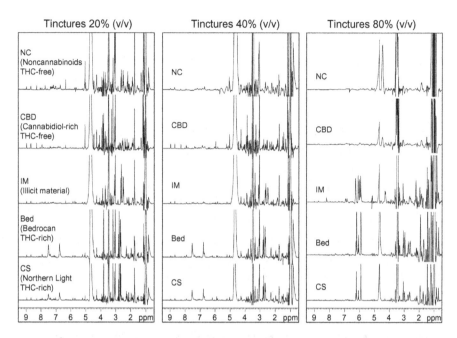

Figure 4.1 ^{1}H-NMR experiments (500 MHz) with suppression of the ethanol and water signals. Comparison of three tinctures (20%, 40%, and 80%, v/v) from five different cannabis cultivars (CS = Northern Lights 5 crossed with Haze, THC-rich; Bed = Bedrocan, THC-rich; IM = illicit material; CBD = cannabidiol rich, THC-free; NC = noncannabinoids, THC-free). In the 80% tinctures, the typical cannabinoids signals appear around 6 ppm in the spectra of the THC-rich plants (CS and Bed) and the illicit material (IM). While these signals in the CS and Bed tinctures are similar, the IM tincture presents a different profile for these cannabinoids signals as well as other signals at 4.35, 6.90, and 8.22 ppm that are unique to this tincture. Both THC-free cultivars (CBD and NC) do not show these cannabinoid signals within the spectra and they are clearly distinguishable from the other tinctures. In the 40% tinctures from both THC-rich plants, cannabinoids signals appear greatly reduced around 6 ppm while other aromatic signals appear in the spectra at 7.45 and 6.75 ppm. These are not detected in any of the other 40% tinctures and they represent a marker for the THC-rich plants in this analysis. Other minor protons are observed between 5 and 9 ppm for all the 40% tinctures and they can be used to compare in detail the various samples. For the 20% tinctures a set of proton signals around 2.55 ppm is characteristic of the IM and NC samples, while a similar set of protons is detected around 2.75 ppm only for CS and Bed samples. Source: From Ref. 27.

extracts and a rationale for use of the method in developing chemical specification for pharmaceutical purposes was made. In this instance, the most selective signals in the aromatic and aliphatic area for the selected cannabis constituents Δ^9-THC, Δ^9-THCA, CBD, CBDA, CBG, and CBGA were identified, and a pattern of a few key proton signals was compiled for fast identification without the need to analyze the complete proton signal set of complex mixtures.[10] This relatively targeted approach can be contrasted with more comprehensive and nontargeted NMR approaches for complex mixtures when combined with multivariate pattern recognition and principal component analysis

(PCA). When performed in parallel, the targeted analysis of key chemicals or signals that can be attributed to a compound or compound class in a sample can amend PCA results and be particularly useful in quality control assessment because it yields information in addition to the mathematical data output of PCA.[10]

SEPARATION AND ANALYSIS OF CANNABIS USING GAS CHROMATOGRAPHY

More extensive chemical characterization of cannabis and cannabis-derived drug substances typically requires the use of several carefully chosen and applied analytical methods, including high-resolution chromatographic separation with either broad spectrum or targeted detection. Capillary gas chromatography (GC) allows high-resolution separation of volatile constituents and can be coupled to several compatible modes of detection, including the universal thermal conductivity detector or the more sensitive and selective flame-ionization (FID), photoionization, electrochemical, nitrogen-phosphorous, and mass selective detectors. Because of their sensitivity to cannabinoids and other chemicals of interest, GC-FID[28−33] and GC-MS are commonly used. However, the high temperatures required for the injection port to convert the liquid extract into the gas phase for gas−liquid phase chromatographic separation results in the rapid decarboxylation of Δ^9-THCA and CBDA and the thermal degradation of other chemical constituents, which needs to be properly accounted for in the analysis of cannabis and cannabis-derived drug products. As a result of the rapid decarboxylation of these aromatic carboxylic acids, the products are retained and elute from the chromatographic separation with the preexisting neutrals (ie, as the total amount of free + acid equivalents). Derivatization of Δ^9-THCA, CBDA, and other phytocannabinoids containing carboxylic acids to their trimethylsilyl esters before injection and analysis using GC is one approach that can stabilize these derivatized forms and allow the differential separation and detection of the acid and free forms. Alternative analytical approaches such as LC-based separations or NMR spectroscopy can detect and quantitate the chemical constituents without decarboxylation of the aromatic carboxylic acids.

SEPARATION AND ANALYSIS OF CANNABIS BY THIN LAYER, LIQUID, AND CONVERGENCE CHROMATOGRAPHY

LC methods that have been used to characterize cannabis also vary considerably across their approaches to separation and detection. Thin layer and flash chromatography continue to be used for their cost-effective resolution of cannabinoid-containing mixtures.[34,35] More frequently, analytical separation is done using reverse-phase high-performance or ultra-performance LC. In addition, the combination of high-pressure liquid CO_2 and high-resolution LC columns packed with small particles ($<$2.1 μM) has enabled the refinement of supercritical fluid chromatography and the development of ultra-performance convergence chromatography (UPC2) systems. This technique provides a mode of analytical or preparative chromatography that is relatively low in cost and environmentally friendly and allows rapid resolution of a wide range of nonpolar and low polarity compounds.[36] UPC2 is a unique, orthogonal mode of separation for the characterization of phytocannabinoids and other constituents in cannabis dosage forms as compared to reverse-phase UPLC (Fig. 4.2). The utility of supercritical fluids in analytical and preparative scale applications in medicinal cannabis is increasingly recognized.[37] Both UPLC and UPC2 are compatible with ultraviolet adsorption, diode array detection, and MS, combinations that are particularly well-suited for the characterization of pharmaceuticals because of their sensitivity for detection and quantitation. These two approaches can accomplish the separation and characterization of nonvolatile or thermally labile substances such as the aromatic acid and neutral forms of phytocannabinoids in the absence of the derivatization step that is required in GC (illustrated in Fig. 4.2).

BROAD-SPECTRUM CHEMICAL PROFILING

Comprehensive broad-spectrum chemical fingerprinting of constituents using chromatography-coupled MS has become increasingly useful for studying plant biochemistry, establishing chemotaxonomy classifications, and providing information for quality control of medicinal plants.[12] Several popular MS platforms enable high-resolution GC and/or LC separation of the chemicals, efficient ionization of the molecules as they elute, and sensitive detection of the ions and their fragment or adduct ions according to their mass-to-charge ratio (m/z).

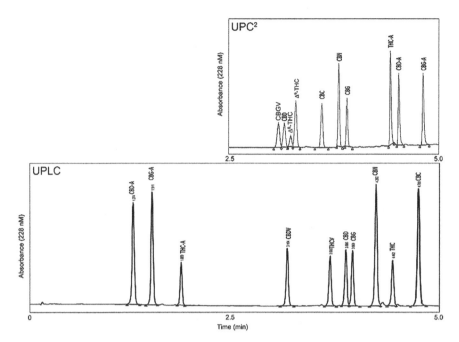

Figure 4.2 Comparison of phytocannabinoid separation using UPLC (top) and UPC² (bottom). Note the differing orders of retention that occur using the two separation systems. The UPLC conditions involved an Acquity CSH C18 2.1 × 50 mm 1.7 µm column maintained at 40°C. Mobile phase A consisted of 10 mM ammonium bicarbonate (pH 10.0) and mobile phase B was acetonitrile. The total flow rate was held at 1.00 mL/min. The initial mobile phase composition was held at 64% A:36% B for 0.5 min, then linearly ramped over 5 min to a final composition of 16% A:84% B. Detection was UV absorbance at 228 nm. The UPC² conditions were programmed to maintain the pressure at 1500 psi and the column at 60°C. The separation was performed at a flow rate of 1.7 mL/min using a concave gradient (curve 8) starting at 1.7 min from the initial conditions of 94.5% CO₂ (mobile phase A) and 5.5% 80:20 isopropanol:acetonitrile with 2% water (mobile phase B) and ending at 70% A and 30% B at 6.8 min. The analytical column was a UPC2 BEH 2-EP (3 × 150 mm, 1.7 µM particle size). Detection was at 228 nM.

These systems may differ in their resolution, ionization modes, efficiencies, and whether they produce adducts or fragmentation ions that spread a molecule's m/z signal across additional ion currents, leading to decreased sensitivity. However, adduct ions and/or fragmentation can give valuable structural information, and both are advantageous in the identification of unknown components with MS-MS technologies or mass spectral library searching. Because of the sensitivity and rapid duty cycle of modern mass spectrometers, a complex sample extract can usually be prepared for analysis, separated with high-resolution chromatography, and subjected to detailed mass spectral data acquisition strategies in less than an hour. Samples typically include internal standards and, in the case of high-resolution MS, continuous mass-axis calibration to facilitate quantitation, mass accuracy, and comparisons between samples. In addition, due to inherent variation in chromatographic retention times and ionization efficiencies among samples, software-based signal processing and deconvolution approaches are used for the alignment of chromatographic peaks, automated baseline correction and background subtraction, and alignment and integration of all component peaks across all samples. Subsequent statistics and bioinformatic tools can then interrogate the data and provide a detailed view of the differences and similarities between samples.[12]

Chromatography and MS-based analytical techniques have been used in studies of cannabis pollen,[11] cannabis seedlings,[6] and mature cannabis inflorescence. In many instances, a characteristic chemical profile signature able to differentiate distinct chemotypes can be identified. In 28-day-old cannabis seedlings, a GC-MS method combined with chemoinformatic classification using a support vector network with supervised learning of the relative proportions of guaiol, γ-eudesmol, bulnesol, α-bisabolol, THV, CBD, Δ^9-THC, and CBN was used to discriminate high Δ^9-THC "drug-type" from low Δ^9-THC "fiber-type" cannabis, with low false-positive rates.[6] This approach was investigated as an alternative to the time-consuming process required for growth of seeds or seedlings to maturation before analysis (eg, as might be required for classification by law enforcement agencies). The positive discrimination in seedlings likely reflects the observations from a previous analytical study showing that, 28 days after sowing, the CBD/Δ^9-THC and CBG/CBD ratios in cannabis leaves are relatively constant and consistent with the ratios determined on mature inflorescences.[16] Broad-spectrum profiling of chemical constituents in cannabis combined with PCA has also been investigated as a means to differentiate between specific drug-type cannabis chemotypes.[1,7,8,15] In one such study, 11 cannabis varieties were grown under the same environmental conditions. Samples of mature plant material were assayed using GC-FID for quantitative analysis of cannabis monoterpenoids, sesquiterpenoids, and cannabinoids. In total, 36 compounds were identified and quantified in the 11 varieties, and PCA of the quantitative data enabled each variety to be chemically discriminated from most of the other varieties tested (Fig. 4.3). This example illustrates how the approach may be useful in guiding pharmacological or clinical studies to examine the potential

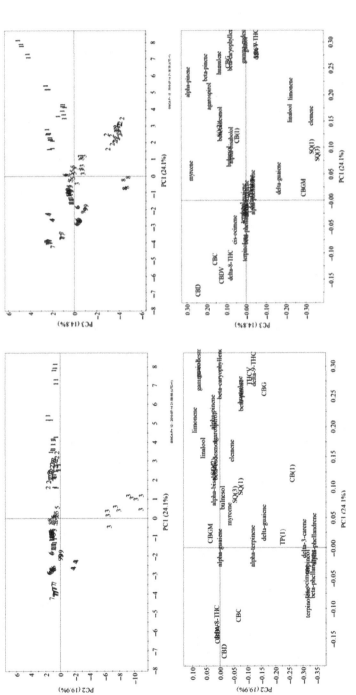

Figure 4.3 Principal components analysis of 36 chemical constituents detected using gas chromatography–flame-ionization detection in 11 cannabis varieties. PC1 versus PC2 (left panels) and PC1 versus PC3 (right panels), presented as factor score scatter plots (top) and loading plots (bottom). The 11 varieties of cannabis are identified by numbers (1) AO, (2) Bedropuur, (3) Bedrocan, (4) Bediol, (5) AG, (6) AE, (7) Ai94, (8) AN, (9) AF, (10) AM, and (11) AD. Source: From Ref. 7.

interactions of the volatile chemical constituents of cannabis and their therapeutic effectiveness. This method could be implemented in the quality control and contribute to the release specifications defined for a particular medicinal cannabis drug substance or dosage formulation. However, further improvement and testing is needed, as a similar study had difficulty discriminating drug-type cannabis accessions from one another.[15] Moreover, Hillig[15] concluded that sesquiterpenoids were more important then monoterpenoids in chemically differentiating cannabis varieties, whereas Fishedick et al.[7] found that monoterpenoids could distinguish between varieties that had similar sesquiterpenoid levels and similar cannabinoid levels. These discrepancies suggest that a combination of genomic and epigenetic approaches with broad-spectrum analytical profiling may be required to further elucidate exactly which biochemical pathways differ in cannabis varieties and how these differences lead to their observed chemical profiles and pharmacological attributes.[7]

TARGETED QUANTITATIVE ANALYTICAL APPROACHES AND COMPENDIAL METHODS

For accurate and reproducible determination of constituents in cannabis, analytical procedures should be developed and validated for the specific drug product or formulation as warranted by its intended use. Once methods have been validated, analytical testing should be rigorously performed against specifications set for release of raw materials, as well as during formulation, processing, and manufacturing, and before final release. A drug product's performance (ie, the delivery of chemicals during use) should also be specified and monitored with validated analytical instruments and methods. For example, smoke or vaporizer yields should be determined on a validated smoking machine with appropriate smoking topography for final release of a batch of cannabis cigarettes or e-cigarette delivery devices. Similarly, disintegration or dissolution testing of cannabis consumables should be used to ensure that products are manufactured with consistent and well-characterized drug delivery. Interestingly, CBD, a nonpsychoactive cannabinoid, was found to be converted to the psychoactive substances Δ^9-THC, 9α-hydroxy-hexahydrocannabinol (9α-OH-HHC), and 8-hydroxy-*iso*-hexahydrocannabinol (8-OH-*iso*-HHC) in artificial gastric juice. This illustrates how the simulation of conditions expected to be encountered with a dose formulation such as CBD-containing cannabis edibles can reveal a potential pharmacological liability associated with a particular route of administration. This type of careful consideration and analytical testing should ideally be built into formulation development before marketing, use, and potential product recall.

The FDA,[20] USP,[21] and other guidance documents for articles of botanical origin and botanical drug development recommend specific procedures that should be used to test cannabis and cannabis-derived drug products. Validated analytical methods for targeted quantitation of selected phytocannabinoid constituents in cannabis have a long history of application in the characterization of illicit material and biological matrices, and the delivery characteristics of cannabis inflorescence during smoking (smoke constituents and quantitative yields) have been extensively studied.[38] The utility of these validated quantitative approaches for targeted constituents in medicinal cannabis and formulations thereof is also increasingly recognized and required. In general, the best available analytical technology should be used to address the issue of analytical resolution and quantitation accuracy. When the resolution is inadequate in one method, multiple methods should be used to provide complementary data for adequate chemical identification and quantification of chemical constituents. For example, when estimates of total Δ^9-THC, CBD, and other phytocannabinoids (aromatic acids and neutrals) are desired, GC-based analysis of solvent extracts that result in rapid decarboxylation of the acidic cannabinoids upon injection can be used to determine "total THC" as a measure of the anticipated strength of the herbal product or formulation.[39-42] However, if quantitation of both acidic and neutral forms of phytocannabinoids is desired, derivatization of the aromatic acids should be performed before GC analysis, or an alternative approach such as HPLC[43-45] or NMR[46] should be adopted or developed and validated. To facilitate method development and validation of these quantitative assays, and enable the identification of components detected in broad-spectrum screening, analytical reference standards and isotopically labeled chemicals are available from chemical suppliers and government programs. For instance the National Institutes of Health has a metabolomics standard synthesis program where investigators can nominate candidate compounds for stable isotope labeling and use in their research.

The analytical testing needed to produce a certificate of analysis for cannabis-derived products or dosage forms should include the assayed concentration of major chemical constituents, excipients, impurities, and degradants over a specified limit, and other determinations such as moisture content and limit tests for foreign organic matter, total ash, acid-insoluble ash, loss on drying, pesticides, microbial and fungal organisms, metals, and residual solvents, as required or warranted by their chemical or biological nature and potential formation during manufacture.[47] All analytical testing methods and results should then be documented and compared against the specifications established for each batch of cannabinoid-containing drug substance or product in a batch production record and a certificate of analysis to fulfill the standards promulgated by current good manufacturing and good laboratory practices. If any analytical procedure performed on a batch fails to meet specifications, an out-of-specification investigation should be conducted so that it can be attributed to an aberration of the measurement process or an aberration of the manufacturing process.

Botanical identification, microscopic examination, and compendial methods used in the characterization of cannabis and cannabis-derived drug products should be well documented and shown to be qualified according to available reference standards before use. As described in the American Herbal Pharmacopeia monograph on cannabis inflorescence, limits that are applicable to herbal materials include solvent and pesticide residues, microbial and fungal limits, and content of certain metals. With the exception of loss on drying and moisture contents of dry cannabis material, the limits are based on general recommendations for botanical ingredients promulgated by national and international agencies as determined through the use of standard pharmacopoeia methods. With cannabis inflorescence, it is recommended that moisture content as determined by weight loss on drying be at or below 15%, preferably around 10%, to prevent microbial or fungal contamination. For other cannabis formulations, the moisture content should be tested for conformance to specifications established based on desired attributes and requirements for chemical content or stability, or product performance characteristics. Other residual organic solvent testing should be done when production or purification processes are known to result in the presence of such solvents. As there is no therapeutic benefit from residual solvents, most residual solvents should be removed to the extent possible to meet product specifications, good manufacturing practices, or other quality-based requirements. There are obvious exceptions for the use of solvates, such as ethanol in tinctures or acetic acids in vinegars; however, the use of certain solvents should be carefully evaluated and justifiable with respect to ensuring the safety of the drug product. Specific limits on certain solvents that can be used in the manufacture of botanical or pharmaceutical products have been established and promulgated by the International Conference on Harmonization (ICH) Q3C, 2011. These solvents are categorized in three classes based on a risk assessment or their potential toxicity level for the consumer or patient. Class 1 solvents include benzene, carbon tetrachloride, 1,2-dichloroethane, 1,1-dichloroethene, and 1,1,1-trichloroethane. These are to be avoided in the manufacture of all herbal or pharmaceutical drug substance, excipients, and drug products. The concentration limits set for class 1 solvents range from 2 parts per million (ppm) for benzene to 1500 ppm for 1,1,1-trichloroethane. Class 2 solvents are associated with less severe toxicity, and should be limited in use or formation during manufacture to protect patients from potential adverse effects. They include acetonitrile, chloroform, hexane, methanol, methylene chloride, tetrahydrofuran, toluene, and other commonly used organic solvents, and their limits are specified individually as both ppm (based on a 10 g daily product dose) and permissible daily exposure. Ideally, less toxic solvents in Class 3 should be used where practical. Class 3 solvents with low toxic potential to man have permissible daily exposure of 50 mg or more per day. They include (among others) acetic acid, acetone, ethanol, ethyl acetate, ethyl ether, formic acid, 1- and 2-butanol, 1-pentanol, and 1- and 2-propanol, which should be limited by GMP or quality specifications of the drug product or justified according to a clearly defined risk/benefit assessment. Most manufacturers and analytical laboratories test for residual solvents by validated headspace GC methods as specified in the USP <467>. However, some Class 2 solvents are not sufficiently volatile to be analyzed by this method and require an alternate validated method to assess residual levels. If only Class 3 solvents are used or suspected to be present, a nonspecific method such as loss on drying may be used.

The analysis of residual pesticides, fungicides, and plant growth regulators and their degradants may best be accomplished by a combination of analytical approaches that provides for the detection of a variety of chemicals, such as that described by Schneider et al.[48] Pesticides and nutrient systems are widely available at hardware stores, specialty indoor hydroponic shops, and online vendors.[49] There are no established application limits for pesticides, and studies have shown that inhalation of cannabis smoke containing pesticides or their chemical residues can occur and may pose significant risks to public health.[49] The heating or combustion of plant material commonly employed for inhalation administration can also produce harmful pyrolysis products of the pesticides,[50] and inhalation exposure to pesticides often causes the most rapid and profound appearance of toxic symptoms (eg, respiratory failure and death from organophosphates). A broad-spectrum approach may also be warranted because of the lack of approved pesticides and product registration for use on cannabis. The American Herbal Pharmacopeia recently published a list of pesticides that are most likely to be used in cannabis cultivation, highlighting 158 active ingredients consisting of acaricides, insecticides, fungicides, and plant growth regulators.[47] This information is supported by anecdotal and analytical sources. In one screen of 50 seized illegal cannabis plants for pesticides using a compendial method for foods of plant origin, samples were analyzed for 160 pesticides with UPLC/MS-MS in positive ESI mode using multiple reaction monitoring and GC-MS in scan mode. Seven pesticides were detected in 19 samples, and five samples contained two pesticides. Four of the pesticides detected were fungicides (propamocarb, tebuconazole, propiconazole, and tolylfluanid); the other three were insecticides or acaricides (imidacloprid, bifenthrin, and hexythiazox).[48] Pesticides and growth regulators are prevalent in medicinal cannabis according to research laboratory and regulatory authority laboratory testing (for review, see Ref. 51). In addition to the previously mentioned QuEChERS approach to screening foods, the Environmental Protection Agency (EPA Residue Analytical Methods[52]) and the FDA (FDA Pesticide Analytical Manual) recommend pesticide testing methods for use on commodity food products that could be applicable to the detection of pesticide residues and their degradants in cannabis-derived drug products.[47] However, even the EPA methods are currently being validated, which likely will not include validation of their use on cannabis-derived drug substances, given their current status as illegal and Schedule I controlled products.

Microbial contamination of natural products such as cannabis can occur even when they are carefully cultivated. Hence, microbial exposure through the use of cannabis or cannabis-derived drug products presents obvious health risks associated with infection, particularly in immune-compromised subjects or individuals with conditions that can be exacerbated by contaminated products. Recommended tolerance limits for unprocessed cannabis (inflorescence, hashish) and processed materials (solid- or liquid-infused edible preparations, oils, topicals, and water processed trichomes/resin) have been proposed by the American Herbal Pharmacopeia: total viable aerobic bacteria $\leq 10^5$, total yeast and mold ≤ 104, total coliforms ≤ 103, bile-tolerant Gram-negative bacteria ≤ 103, and *Escherichia coli* (pathogenic strains) and *Salmonella* spp. not detected in 1 g sample. For CO_2 and solvent-based extracts, these limits are lowered by a log unit, with the exception of the criteria for *E. coli* (pathogenic strains) and *Salmonella* spp., which remain at not detected in 1 g sample. However, it is important to note that significant microbial contamination can occur during postharvesting or postprocessing handling, and that more restrictive limits may be required to be consistent with applicable state, federal, and international regulations, or when used in specific patients. For example, *Aspergillus* spp. in inhaled formulations may be a particular concern. Thus, specific microbial testing should always be employed, and treatments might be warranted to further reduce the microbial risks as much as possible without compromising therapeutic activity.[47] Acceptable methods for microbial testing have been recommended by the FDA in the Bacteriological Analytical Manual.[53]

Cannabis spp. can grow and serve as a bioaccumulator of heavy metals in soil and water, and thus pose a health risk to consumers of cannabis-derived drug substances.[54–58] When grown under carefully controlled conditions, the risk of heavy metal contamination is relatively low. Nevertheless, the American Herbal Products Association has recommended limits for heavy metals in botanical dietary supplements of 10 μg/day for inorganic arsenic, 4.1 μg/day for cadmium, 6.0 μg/day for lead, and 2.0 μg/day for methyl mercury. Limits for heavy metals in drug products and their test methods are also provided in the USP-NF[21] and other international guidance documents. The recommended approaches often include general screening methods as well as specific tests for individual elements and their species (oxidation states) of concern such as aluminum, arsenic, iron, lead, mercury, and selenium. As recommended in the USP-NF, two instrumental procedures are considered appropriate for elemental impurities: inductively coupled plasma–atomic (optical) emission spectroscopy

(ICP-AES or ICP-OES) and inductively coupled plasma—mass spectrometry (ICP-MS). Both procedures must be validated for the specific instrumentation and sample preparation and analysis to ensure suitability for intended use. The level of validation necessary to ensure that a procedure is appropriate for its intended purpose is dependent upon whether a limit test or a quantitative determination is necessary.[21]

REFERENCES

1. Hazekamp A, Fischedick JT. Cannabis—from cultivar to chemovar. *Drug Test Anal.* 2012;4 (7–8):660–667.

2. Potter DJ. Cannabis horticulture. In: Pertwee RG, ed. *Handbook of Cannabis*. New York, NY: Oxford University Press; 2014:65–88.

3. Potter DJ. A review of the cultivation and processing of cannabis (*Cannabis sativa* L.) for production of prescription medicines in the UK. *Drug Test Anal.* 2014;6(1–2):31–38.

4. Stott CG, Guy GW. Cannabinoids for the pharmaceutical industry. *Euphytica.* 2004;140 (1–2):83–93.

5. Hazekamp A, Pappas G. Self-medication with Cannabis. In: Pertwee RG, ed. *Handbook of Cannabis*. New York, NY: Oxford University Press; 2014:319–338.

6. Broséus J, Anglada F, Esseiva P. The differentiation of fibre- and drug type Cannabis seedlings by gas chromatography/mass spectrometry and chemometric tools. *Forensic Sci Int.* 2010;200(1–3):87–92.

7. Fischedick JT, Hazekamp A, Erkelens T, Choi YH, Verpoorte R. Metabolic fingerprinting of *Cannabis sativa* L., cannabinoids and terpenoids for chemotaxonomic and drug standardization purposes. *Phytochemistry.* 2010;71(17–18):2058–2073.

8. Ilias Y, Rudaz S, Mathieu P, Christen P, Veuthey JL. Extraction and analysis of different Cannabis samples by headspace solid-phase microextraction combined with gas chromatography-mass spectrometry. *J Sep Sci.* 2005;28(17):2293–2300.

9. Omar J, Olivares M, Amigo JM, Etxebarria N. Resolution of co-eluting compounds of *Cannabis sativa* in comprehensive two-dimensional gas chromatography/mass spectrometry detection with multivariate curve resolution-alternating least squares. *Talanta.* 2014;121:273–280.

10. Peschel W, Politi M. ^{1}H NMR and HPLC/DAD for *Cannabis sativa* L. chemotype distinction, extract profiling and specification. *Talanta.* 2015;140:150–165.

11. Rothschild M, Bergstrom G, Wangberg SA. *Cannabis sativa*: volatile compounds from pollen and entire male and female plants of two variants, Northern Lights and Hawaian Indica. *Bot J Linn Soc.* 2005;147(4):387–397.

12. van der Kooy F, Maltese F, Choi YH, Kim HK, Verpoorte R. Quality control of herbal material and phytopharmaceuticals with MS and NMR based metabolic fingerprinting. *Planta Med.* 2009;75(7):763–775.

13. Da Porto C, Decorti D, Natolino A. Separation of aroma compounds from industrial hemp inflorescences (*Cannabis sativa* L.) by supercritical CO_2 extraction and on-line fractionation. *Ind Crop Prod.* 2014;58:99–103.

14. Happyana N, Agnolet S, Muntendam R, Van Dam A, Schneider B, Kayser O. Analysis of cannabinoids in laser-microdissected trichomes of medicinal *Cannabis sativa* using LCMS and cryogenic NMR. *Phytochemistry*. 2013;87:51–59.

15. Hillig KW. A chemotaxonomic analysis of terpenoid variation in Cannabis. *Biochem Syst Ecol*. 2004;32(10):875–891.

16. Pacifico D, Miselli F, Carboni A, Moschella A, Mandolino G. Time course of cannabinoid accumulation and chemotype development during the growth of *Cannabis sativa* L. *Euphytica*. 2008;160(2):231–240.

17. Taschwer M, Schmid MG. Determination of the relative percentage distribution of THCA and Δ9-THC in herbal cannabis seized in Austria—impact of different storage temperatures on stability. *Forensic Sci Int*. 2015;254:167–171.

18. Verhoeckx KC, Korthout HA, van Meeteren-Kreikamp AP, et al. Unheated *Cannabis sativa* extracts and its major compound THC-acid have potential immuno-modulating properties not mediated by CB1 and CB2 receptor coupled pathways. *Int Immunopharmacol*. 2006;6 (4):656–665.

19. Organization WH. *Quality Control Methods for Medicinal Plant Materials*. England: World Health Organization; 1998:214.

20. Administration USFaD. In: Research CfDEa, ed. *Guidance for Industry: Botanical Drug Development (Draft Guidance)*. Silver Spring, MD: Office of Communications, Division of Drug Information; 2015.

21. USP-NFU-N. Rockville, MD: United States Pharmacopeia Convention; 2015.

22. Wianowska D, Dawidowicz AL, Kowalczyk M. Transformations of tetrahydrocannabinol, tetrahydrocannabinolic acid and cannabinol during their extraction from *Cannabis sativa* L. *J Anal Chem*. 2015;70(8):920–925.

23. Díaz-Maroto MC, Pérez-Coello MS, Cabezudo MD. Supercritical carbon dioxide extraction of volatiles from spices: comparison with simultaneous distillation–extraction. *J Chromatogr A*. 2002;947(1):23–29.

24. Omar J, Olivares M, Alzaga M, Etxebarria N. Optimisation and characterisation of marihuana extracts obtained by supercritical fluid extraction and focused ultrasound extraction and retention time locking GC-MS. *J Sep Sci*. 2013;36(8):1397–1404.

25. Da Porto C, Natolino A, Decorti D. Effect of ultrasound pre-treatment of hemp (*Cannabis sativa* L.) seed on supercritical CO_2 extraction of oil. *J Food Sci Technol*. 2015;52 (3):1748–1753.

26. Kauppila TJ, Flink A, Laakkonen UM, Aalberg L, Ketola RA. Direct analysis of cannabis samples by desorption atmospheric pressure photoionization-mass spectrometry. *Drug Test Anal*. 2013;5(3):186–190.

27. Politi M, Peschel W, Wilson N, Zloh M, Prieto JM, Heinrich M. Direct NMR analysis of cannabis water extracts and tinctures and semi-quantitative data on delta9-THC and delta9-THC-acid. *Phytochemistry*. 2008;69(2):562–570.

28. Brenneisen R, ElSohly MA. Chromatographic and spectroscopic profiles of Cannabis of different origins: part I. *J Forensic Sci*. 1988;33(6):1385–1404.

29. Chandra S, Lata H, Mehmedic Z, Khan IA, Elsohly MA. Assessment of cannabinoids content in micropropagated plants of *Cannabis sativa* and their comparison with conventionally propagated plants and mother plant during developmental stages of growth. *Planta Med*. 2010;76(7):743–750.

30. Gambaro V, Dell'Acqua L, Farè F, Froldi R, Saligari E, Tassoni G. Determination of primary active constituents in Cannabis preparations by high-resolution gas chromatography/ flame ionization detection and high-performance liquid chromatography/UV detection. *Anal Chim Acta*. 2002;468(2):245–254.

31. Gröger T, Schäffer M, Pütz M, et al. Application of two-dimensional gas chromatography combined with pixel-based chemometric processing for the chemical profiling of illicit drug samples. *J Chromatogr A*. 2008;1200(1):8−16.

32. Ross SA, Elsohly HN, Elkashoury EA, Elsohly MA. Fatty acids of cannabis seeds. *Phytochem Anal*. 1996;7(6):279−283.

33. Tipparat P, Natakankitkul S, Chamnivikaipong P, Chutiwat S. Characteristics of cannabinoids composition of Cannabis plants grown in Northern Thailand and its forensic application. *Forensic Sci Int*. 2012;215(1−3):164−170.

34. Fischedick JT, Glas R, Hazekamp A, Verpoorte R. A qualitative and quantitative HPTLC densitometry method for the analysis of cannabinoids in *Cannabis sativa* L. *Phytochem Anal: PCA*. 2009;20(5):421−426.

35. Wohlfarth A, Mahler H, Auwarter V. Rapid isolation procedure for Delta9-tetrahydrocannabinolic acid A (THCA) from *Cannabis sativa* using two flash chromatography systems. *J Chromatogr B Analyt Technol Biomed Life Sci*. 2011;879(28):3059−3064.

36. Zhou Q, Gao B, Zhang X, Xu Y, Shi H, Yu L. Chemical profiling of triacylglycerols and diacylglycerols in cow milk fat by ultra-performance convergence chromatography combined with a quadrupole time-of-flight mass spectrometry. *Food Chem*. 2014;143:199−204.

37. Hudalla CJ. Analytical Testing for the cannabis industry: application of ultraperformance convergence chromatography (UPC2). In: *CoSMoS: Conference on Small Molecule Science*; Williamsburg, VA; 2014.

38. Van der Kooy F, Pomahacova B, Verpoorte R. Cannabis smoke condensate I: the effect of different preparation methods on tetrahydrocannabinol levels. *Inhal Toxicol*. 2008; 20(9):801−804.

39. Fairlie K, Fox BL. Rapid, quantitative determination of tetrahydrocannabinol in marihuana by gas chromatography. *J Chromatogr Sci*. 1976;14(7):334−335.

40. Hewavitharana AK, Golding G, Tempany G, King G, Holling N. Quantitative GC-MS analysis of D(9)-tetrahydrocannabinol in fiber hemp varieties. *J Anal Toxicol*. 2005; 29(4):258−261.

41. Barni Comparini I, Centini F. Packed column chromatography, high-resolution gas-chromatography and high pressure liquid chromatography in comparison for the analysis of cannabis constituents. *Forensic Sci Int*. 1983;21(2):129−137.

42. Song CH, Kanter SL, Hollister LE. Extraction and gas chromatographic quantification of tetrahydrocannabinol from marihuana. *Res Commun Chem Pathol Pharmacol*. 1970; 1(3):375−382.

43. Aizpurua-Olaizola O, Omar J, Navarro P, Olivares M, Etxebarria N, Usobiaga A. Identification and quantification of cannabinoids in *Cannabis sativa* L. plants by high performance liquid chromatography-mass spectrometry. *Anal Bioanal Chem*. 2014;406(29):7549−7560.

44. De Backer B, Debrus B, Lebrun P, et al. Innovative development and validation of an HPLC/DAD method for the qualitative and quantitative determination of major cannabinoids in cannabis plant material. *J Chromatogr B Analyt Technol Biomed Life Sci*. 2009;877(32):4115−4124.

45. Smith RN, Vaughan CG. High-pressure liquid chromatography of cannabis. Quantitative analysis of acidic and neutral cannabinoids. *J Chromatogr*. 1976;129:347−354.

46. Hazekamp A, Choi YH, Verpoorte R. Quantitative analysis of cannabinoids from *Cannabis sativa* using 1H-NMR. *Chem Pharm Bull (Tokyo)*. 2004;52(6):718−721.

47. Upton R, Dayu RH, Craker L, et al., eds. *Cannabis inflorescence (Cannabis spp.): Standards of Identify, Analysis, and Quality Control*. Scott's Valley, CA: American Herbal Pharmacopeia, ed; 2013.

48. Schneider S, Bebing R, Dauberschmidt C. Detection of pesticides in seized illegal cannabis plants. *Anal Methods*. 2014;6(2):515–520.

49. Sullivan N, Elzinga S, Raber JC. Determination of pesticide residues in cannabis smoke. *J Toxicol*. 2013;2013:378168.

50. Lorenz W, Bahadir M, Korte F. Thermolysis of pesticide residues during tobacco smoking. *Chemosphere*. 1987;16(2–3):521–522.

51. Stone D. Cannabis, pesticides and conflicting laws: the dilemma for legalized States and implications for public health. *Regul Toxicol Pharmacol: RTP*. 2014;69(3):284–288.

52. United States Environmental Protection Agency. Residue Analytical Methods. <http://www2.epa.gov/pesticide-analytical-methods/analytical-methods-measuring-pesticide-residues>; Accessed 14.10.15.

53. Administration USFaD. Bacteriological Analytical Manual. <http://www.fda.gov/Food/FoodScienceResearch/LaboratoryMethods/ucm2006949.htm>; Accessed 14.10.15.

54. Angelova V, Ivanova R, Delibaltova V, Ivanov K. Bio-accumulation and distribution of heavy metals in fibre crops (flax, cotton and hemp). *Ind Crop Prod*. 2004;19(3):197–205.

55. Khan MA, Wajid A, Noor S, Khattak FK, Akhter S, Rahman IU. Effect of soil contamination on some heavy metals content of *Cannabis sativa*. *J Chem Soc Pak*. 2008;30(6):805–809.

56. Shi G, Liu C, Cui M, Ma Y, Cai Q. Cadmium tolerance and bioaccumulation of 18 hemp accessions. *Appl Biochem Biotechnol*. 2012;168(1):163–173.

57. Zavada J, Bouchal T, Fryc J, Nadkanska H, Blazek V. *Cannabis sativa* as a biosorbent for the removal of heavy metals from aqueous solutions. In: *Paper presented at: International Multidisciplinary Scientific GeoConference Surveying Geology and Mining Ecology Management, SGEM2013*.

58. Zerihun A, Chandravanshi BS, Debebe A, Mehari B. Levels of selected metals in leaves of *Cannabis sativa* L. cultivated in Ethiopia. *SpringerPlus*. 2015;4(1).

CHAPTER 5

Quality Control and Stability Assessment

CHALLENGES IN QUALITY CONTROL AND SAFETY OF CANNABIS AND CANNABIS-DERIVED DRUGS

Medical cannabis products that are obtained from dispensaries or state programs may follow specific international, national, or state requirements related to growing, formulation, manufacturing, marketing, and distribution. However, in general the standards for these products and their labeling have not been thoroughly researched or harmonized. Cannabis dispensaries in the United States are providing their users with products that have not been reviewed or approved by the FDA as mandated by the Compassionate Investigational New Drug Program. These dispensaries are also not legally registered or licensed by the DEA to distribute cannabis or cannabis-derived materials, and the distribution of these substances may not be well-documented or controlled as is the case with pharmaceutical substances regulated by the FDA or DEA. Formulations provided by dispensaries vary widely in nature and origin of materials, with many cannabis herbal chemotypes processed in different ways. There are myriad solid and liquid products for various methods of inhalation as smoke or vapor, ingestion, and delivery to mucous membranes. Labeling practices vary between states and even dispensaries, often providing the user with limited information on ingredients. Batch production records for raw materials and formulations are generally not available. Varieties of cannabis are often distinguished by popular names that can vary from place to place. Fig. 3.1 gives an idea of the nebulous labeling. While the content of Δ^9-tetrahydrocannabinol (Δ^9-THC) and other phytocannabinoids is often provided on the label or packaging, the nomenclature used to describe the chemical constituents can be unclear or lacking in sufficient detail. For example, the decarboxylated forms of Δ^9-THC carboxylic acid (Δ^9-THCA) and cannabidiol carboxylic acid (CBDA), Δ^9-THC and cannabidiol (CBD), respectively, are often referred to as "activated cannabinoids" on edible products.

The Analytical Chemistry of Cannabis. DOI: http://dx.doi.org/10.1016/B978-0-12-804646-3.00005-9

VARIABILITY IN COMPOSITION AND STRENGTH

As with most natural product-derived formulations, cannabis tinctures and extracts are associated with variable pharmacological composition and strength, leading to difficulty in obtaining replicable therapeutic effects. This inconsistency has generally been attributed to the varying origins and age of the plant material and variations in preparation.[1] With some notable exceptions (eg, Sativex), this historical trend appears to be continuing in today's medicinal cannabis dispensaries, where solvent and compressed gas extracts containing concentrated phytocannabinoids are manufactured with varying levels of analytical assessment, quality control, and release specifications, are often marketed as hash oil, butane hash oil, dabs, shatter, wax, and other popular names, and can vary in chemical content and strength between suppliers or even between batches from the same supplier. In addition, results from stability assessments of concentrated cannabis extracts reveal that a steady decay of phytocannabinoids, including Δ^9-THCA and CBDA, can occur during storage.[2] Specifically, the degradation of Δ^9-THCA in samples exposed to light at 22°C is more pronounced than in samples stored in darkness at 4°C. Sensitivity to light has also been observed in other solvents and samples of phytocannabinoids. As a primary chemical degradant of Δ^9-THCA is cannabinol carboxylic acid (CBNA), the concentration of CBNA and cannabinol (CBN) also increases during storage, and the observed increase is more pronounced for samples exposed to light at 22°C than those stored in darkness at 4°C. The concentration of CBDA in extracts decreases during storage, especially for samples exposed to light at 22°C, which is consistent with cyclization of CBDA to Δ^9-THCA, followed by the degradation of Δ^9-THCA to CBNA.[2] While CBDA and CBD are generally thought of as nonpsychoactive, CBD can readily cyclize to Δ^9-THC under acidic conditions that can be achieved using relatively simple and routine synthetic chemistry reagents and methods,[3,4] and this conversion has also been shown in laboratory simulations of the acids found in the human stomach.[5] Humans in some studies have been given relatively high oral doses of CBD,[6,7] but it remains to be clearly determined whether the conversion of CBD to active Δ^9-THC degradants occurs in vitro after acute or chronic dosing, and whether this can produce CB1 and CB2 receptor agonist-like effects in man.

CONTENT AND LABELING INACCURACIES AND VIOLATIONS

Particularly concerning are the recent observations that errors or inaccuracies in chemical content and labeling are prevalent in products purchased from medical cannabis markets in the United States. In one instance, over 75 products were purchased (47 brands) and analyzed. The investigators (see Vandrey Close-Up: Cannabis Quality Assurance?) reported that only 17% were accurately labeled, 23% were underlabeled, and 60% were overlabeled with respect to THC content.[8] As a clear indication of the increasing concern over quality control in the medical marijuana market in the United States, the FDA reported in February 2015 that there were firms marketing unapproved cannabidiol-containing drugs *"for the diagnosis, cure, mitigation, treatment, or prevention of diseases. Some of these firms claim that their products contain cannabidiol (CBD). FDA has tested those products and, in some of them, did not detect any CBD. It is important to note that these products are not approved by FDA for the diagnosis, cure, mitigation, treatment, or prevention of any disease, and often they do not even contain the ingredients found on the label. Consumers should beware purchasing and using any such products."* Excerpted language used in one warning letter further illustrates the lack of quality control and approval of these drug products, their chemical constituents, and their medical claims and labeling as *"'new drugs' under section 201(p) of the Act [21 U.S.C. § 321(p)] because they are not generally recognized as safe and effective for use under the conditions prescribed, recommended, or suggested in the labeling. New drugs may not be legally introduced or delivered for introduction into interstate commerce without prior approval from the FDA, as described in section 505(a) of the Act [21 U.S.C. § 355(a)]; see also section 301(d) of the Act [21 U.S.C. § 331(d)]. FDA approves a new drug on the basis of scientific data submitted by a drug sponsor to demonstrate that the drug is safe and effective."* and noted that the products were also *"misbranded within the meaning of section 502(f)(1) of the Act [21 U.S.C. § 352(f)(1)], because the labeling for these products fails to bear adequate directions for use. "Adequate directions for use" means directions under which a layman can use a drug safely and for the purposes for which it is intended [21 CFR § 201.5]. Your marketed products are offered for conditions that are not amenable to self-diagnosis and treatment by individuals who are not medical practitioners; therefore, adequate directions for use cannot be written so that a layperson can use this drug safely for its intended purposes. FDA-approved drugs which bear their FDA-approved labeling are exempt from the requirements that they bear adequate directions for use by a layperson [21 CFR §§ 201.100(c)(2) and 201.115]. Because the above-mentioned products lack FDA-approved applications, they are not exempt under 21 CFR §§ 201.100(c)(2) and 201.115. For these reasons, these products are misbranded under section 502(f)(1) of the Act. The introduction or delivery for introduction of a misbranded drug into interstate commerce is a violation of section 301 (a) of the Act [21 U.S.C. § 331(a)].*"[9]

FOODS AND PHARMACEUTICALS

Beyond content and labeling concerns, MacCoun and Mello have noted that edibles resembling sugary snacks pose several clear risks, including inadvertent consumption and overintoxication.[10] Indeed, these products were often found to be distributed without safety packaging, tamper resistance, or clear unit doses, and inadvertently consumed by both children and adults who became intoxicated or suffered adverse effects.[11,12] Furthermore, a single candy could constitute a single serving several times the phytocannabinoid content that is considered a safe dose.[10] The dangers of overdosing are exacerbated by the absence of appropriate labeling and directions for use, as well as the lack of an FDA-approved antagonist to rapidly reverse the adverse effects. In response to repeated incidences of inadvertent consumption, Colorado and Washington state have adopted regulations for edibles requiring child-resistant packaging, a warning to "keep out of the reach of children," and labeling describing a standard serving size, along with prohibition of packaging and advertising that target children. However, neither state requires warnings that ingested marijuana can have different effects from smoked marijuana, or packaging that is clearly distinguishable from ordinary food products.[10] Furthermore, each nonphytocannabinoid ingredient in a cannabis edible or ingestible product could be considered the equivalent of an excipient in a pharmaceutical dosage formulation. As these ingredients can profoundly affect the properties of the dosage formulation, they should be carefully considered and tested for compatibility and stability, and fully documented and controlled during the manufacturing and labeling process.

BEST PRACTICES AND QUALITY CONTROL

Good Agricultural and Collection Practices

Cannabis contains a wide range of chemical constituents and can be selected and grown under various conditions that can affect their presence and strength, with obvious implications for pharmacological effects. Because of the inherent complexity and variability of cannabis and the limitations in the capacity for analytical techniques to characterize all constituents solely by chemical or biological means, the reproducibility of the raw botanical material for medical or veterinary use should be maximized by using good agricultural and collection practices (GACP) to the greatest extent possible in all aspects of growing, harvesting, drying, and storage (for GACP guidance, see Refs. 13 and 14). As described by Potter, and by Stott and Guy, the quality and reliability of cannabis used as a drug substance or in the preparation of cannabis-containing dosage formulations should be ensured by specification and rigorous control of the seeds or rooted clones used and the growth conditions employed, which should be described in a detailed protocol or batch production record.[15-17] Another important aspect for production of medicinal cannabis plants or derived substances is the use of hygienic procedures to minimize microbiological load. This is particularly true for indoor growing, where systematic control of soil conditions, water, nutrients, temperature, and lighting is possible and advantageous for maximizing yields, consistency in the content of chemical constituents, and the quality and reproducibility of the final product. After the cannabis plants are harvested, they should be carefully processed and dried in the dark in a ventilated area. Prompt drying is required to minimize bacterial and fungal decay. The flowering tops (inflorescence) of the female plants are subsequently removed and manicured to retain their phytocannabinoid content and strength while eliminating stems and unwanted materials. A final moisture content at or slightly below 10% (w/w) is generally considered suitable for long-term storage and use.[18] Specific methods for harvesting, drying, manicuring, and storing the inflorescence may be allowed to vary between products but should be documented and adhered to, as this can have a significant impact on the consistency of a final product's chemical constituents. Long-term stability of the phytocannabinoids and chemical constituents in harvested and manicured cannabis is best insured through the use of secure, environmentally controlled storage conditions that eliminate exposure to air (moisture), light, and elevated temperatures.[19] However, even under ideal conditions, the cannabinoid acids (eg, Δ^9-THCA and CBDA) undergo slow degradation to their decarboxylated forms (Δ^9-THC and CBD, respectively), and the terpenes and flavonoids can also isomerize, degrade, or suffer loss through evaporation.[18,19] While aging or further processing of cannabis may be desired, maintenance of the original content and stability of the chemical constituents is likely to be best achieved by using cold temperature storage under an inert atmosphere in the absence of light.[20]

Good Manufacturing Practices

The manufacturing process is also a critical activity where quality control and best practices are required to ensure the suitability of medicinal products. Many guidance documents are available for reference and use in the development of botanical medicines, including the World Health Organization's (WHO) guidelines on good manufacturing practices (GMP) for herbal medicines[21] and the US FDA Guidance for Industry: Botanical Drug Development (Draft Guidance).[22] Any processing, manufacturing, packaging, and storage of raw cannabis and cannabis-derived chemicals for medicinal use should be performed and documented in accordance with GMP and other applicable regulations and best practices. For example, the use of generally recognized as safe ingredients is preferred; as is the use of appropriately cleaned and approved sample handling, processing, and packaging materials. Additional consideration should be given to the final dosage formulation's chemical constituents, the route of administration, and other matters with implications for use and safety. This has been a particular concern with the increasing use of vaporizers containing flavorings and additives that are generally accepted as safe when consumed orally but have not been rigorously evaluated for safety with routes of administration that involve combustion or heating and inhalation.

Good Laboratory Practices

When targeted analytical methods are used for quantitative determination of specific ingredients, they should be validated for specificity, linearity, accuracy, precision, and ruggedness, and only then used to assess the product's constituents against specifications. The FDA and other regulatory organizations provide guidance documents that represent the current thinking on method validation and other subjects related to good laboratory practices (GLP) to assist industry in satisfying applicable statutes and regulations. For analytical method validation, relevant literature and applicable guidance from a regulatory perspective include the following:

- USP-NF, Validation of Compendial Methods; USP Pharmacopeia 35, United States Pharmacopeial Convention, Inc., Rockville MD, May 1, 2012–December 1, 2012.
- U.S. FDA, Center for Drug Evaluation and Research (CDER), Reviewer Guidance on Validation of Chromatographic Methods, November 1994.
- U.S. FDA, CDER, ICH Guideline for Industry, Text on Validation of Analytical Procedures. ICH-Q2A, March 1995.
- U.S. FDA, CDER, Validation of Analytical Procedures: Methodology. ICH-Q2B, November 1996.
- U.S. FDA, CDER, Draft Guidance for Industry, Analytical Procedures and Methods Validation CMC Documentation, August 2000.

Validation or qualification of instrumentation involves tasks whereby the installation, operation, and performance of the instrument is verified by a series of tests to determine that it performs and provides results that are valid and accurate. Following successful validation of an assay procedure, an analytical method is documented in a manner consistent with Chemistry, Manufacturing and Controls (CMC) submission to the FDA as part of an Investigational New Drug (IND) or New Drug Application (NDA). Although a summary of validation results is included in the analytical method for information purposes, the details of validation are documented in a separate validation report. System suitability specifications are based on the entire validation package and include acceptance criteria for capacity factor (k'), tailing factor (T), and theoretical plate number (N). If multiple analytes are involved, limits for resolution (RS) and relative retention may also be included. System suitability testing is routinely done in advance of sample analysis to provide documentation that the system performance, and hence data quality, is adequate to the intended purpose.

With a botanical drug product or derived substance, the types of impurities present may range from structurally similar organic compounds (eg, biosynthetic intermediates, degradation products) to unexpected byproducts or contaminants, pesticide residues, metals, residual solvents, and water. Although no single analytical technique is capable of detecting all possible types of impurities, information from a variety of analytical, physical, and chemical characterization techniques must be considered in combination to arrive at a reasonable estimate of absolute drug purity in the absence of a defined reference standard. In some instances, such as the determination of noncannabinoid terpene profiles, broad-spectrum chemical profiling methods may be used to illustrate consistency or discrepancies between batches of drug substance or formulated product. Similarly, the phytocannabinoid constituents of primary interest such as Δ^9-THCA, Δ^9-THC, Δ^8-THCA, Δ^8-THC, CBDA, CBD, CBNA, CBN, cannabigerolic acid (CBGA), cannabigerol (CBG), cannabigerovarinic acid (CBGVA), cannabigerovarin (CBGV), cannabichromenic acid (CBCA), cannabichromene (CBC), tetrahydrocannabivarinic acid (Δ^9-THCVA), and tetrahydrocannabivarin (Δ^9-THCV) should be profiled when applicable, while the principal phytocannabinoid constituents should be quantitated using validated chromatographic, spectrometric, or spectrophotometric methods.

RELEASE TESTING AND CHARACTERIZATION OF CHEMICAL DELIVERY

Finished drug products require release testing that may include appearance, identity, assay, content uniformity, average weight and weight variation, dissolution, moisture content, and related substances (eg, impurities and degradation products) as per established specifications. The postformulation tests performed on a batch of dosage forms depend on the nature of the formulation. Packaged solutions are typically tested for clarity, pH, weight or volume variation, content uniformity, and intact container-closure seals. Other specialized tests, including the USP sterility test and pyrogen test, are done on injectable solutions. Tablets and capsules are typically tested for weight variation and content uniformity (assay), as well as disintegration characterization or dissolution by using a USP-approved apparatus to measure the release of chemical ingredients over time. By extension, such characterization might be warranted for all ingestible forms of cannabis-derived drug products so that the manufacturer and the consumer would know how products differed in their chemical stability, compatibility, and release characteristics. Inhaled drug devices such as nasal sprays, metered-dose inhalers, dry powder inhalers, vaporizers and nebulizers also require careful analytical characterization of chemical contents and delivery characteristics. For example, the delivery of vaporized (noncombusted) cannabis or phytocannabinoid concentrates from e-cigarettes and other vaporization delivery devices has been shown to reduce harmful exposure as compared to combustion,[23] but this exposure differential can vary depending upon vaporization temperature,[24] the mechanical properties of the device, and the chemical ingredients and excipients.[25] The use of nonpharmaceutically tested vehicles and flavorings found in nonregulated products poses additional risks for chemical reactions, pharmacokinetic interactions, and other product liabilities. Thus, while vaporization is generally assumed to be considerably safer than smoking, the unknowns involved in the development and use of novel drug formulations and delivery devices may still represent a significant public health concern.[26] This is particularly true because of their potential for easing or facilitating initiation and use, altering sensory perceptions, serving as sensory cues associated with drug administration, and generally increasing their abuse or dependence liability.[27] For cigarettes, e-cigarettes, vaporizers, nebulizers, and other inhalation devices, chemical delivery should be characterized with a validated apparatus to produce and collect smoke or vapor and the results included on the certificate of analysis and product specifications. This approach should also allow determination of chemical constituents, particle size, and particle deposition characteristics using a cascade impactor or similar device, as particle size has been shown to influence the deposition, retention, and absorption of constituents as they are inhaled.[28–30] Release testing data for a batch are reviewed to verify conformity, and then released upon satisfactory conformance with the specifications. Any unexplained deviations should be investigated immediately and the results documented to inform the decision-making process regarding approval and release or rejection of the batch.

STABILITY ASSESSMENT

Stability studies also play an important role in the development of cannabis and cannabis-derived drug formulations for use in humans and companion animals. In this context, the purpose of stability testing is to provide evidence on how the quality of a drug substance or drug product varies with time under the influence of a variety of environmental factors, such as temperature, humidity, and light, and enables recommended storage conditions, retest periods, and the shelf-life to be established. A considerable amount of information has already been obtained on the stability of selected phytocannabinoid chemical standards in pharmaceutical vehicles and solvents under various environmental storage conditions,[31−34] and the stability of cannabis in bulk and in cigarettes when stored under various environmental conditions has also been defined over extended periods (Fig. 5.1).[35−38] For instance, it has been reported that Δ^9-THCA degrades exponentially via decarboxylation, with concentration half-lives of approximately 330 and 462 days in daylight and darkness, respectively. In this study, the degradation of neutral Δ^9-THC occurred somewhat slower. This study also demonstrated that when cannabinoids were stored in extracted form at room temperature the degradation rate of acidic Δ^9-THC increased significantly relative to resin material, with concentration half-lives of 35 and 91 days in daylight and darkness, respectively.[39] These data suggest that cannabinoid-containing products should be maintained under refrigerator or freezer conditions in the absence of light to best ensure that they retain quality. Each drug product needs to have stability determined in accordance with the FDA and ICH guidelines (Fig. 5.1).

Before executing formal stability studies, stability-indicating analytical methods should be developed and validated, and preliminary information on the inherent stability of the bulk drug should be gained through stress testing. Stress testing is usually done on a single batch of drug substance in the solid-state form and in solution, under stress conditions of elevated temperature, high humidity, light, and atmospheric oxygen. In planning stability studies on bulk drug substances or drug formulations, it is important that the storage conditions be carefully chosen so that appropriate data are gathered to provide early indications of potential stability problems and to support tentative expiration dating. Many factors can affect the stability of a drug substance, including the storage time, temperature, humidity, container/closure system, headspace atmosphere, and inclusion or exclusion of light. Stability testing is planned and conducted according to the FDA guidance so as to cover those physical and chemical properties susceptible to change during storage and likely to influence the quality, safety, or efficacy of the drug material:

- U.S. FDA, CDER, ICH Guidance for Industry, Q1A(R2) Stability Testing of New Drug Substances and Products, Revision 2, November 2003
- U.S. FDA, CDER, ICH Guidance for Industry, Impurities in New Drug Substances, Revision 2, June 2008.

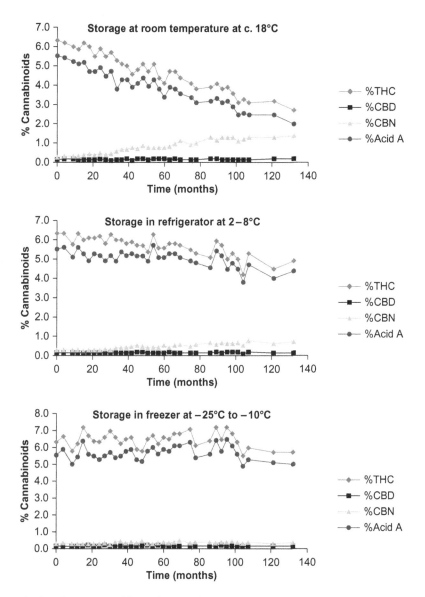

Figure 5.1 Cannabis cigarette stability study.

Photostability testing is an integral part of stress testing for a drug substance and should be done on most cannabinoid-containing drug products. The objective is to determine whether alternative formulations, light-resistant packaging, or special labeling is needed to reduce exposure to light.

The FDA and ICH guidance document for the design, conduct, and evaluation of photostability studies is:

• U.S. FDA, CDER, ICH Guidance for Industry, Q1B Photostability Testing of New Drug Substances and Products, November 1996.

The results must be verified through long-term testing or actual shelf-life monitoring. Shelf-life monitoring samples are stored according to their intended conditions and are assayed periodically to determine integrity and strength. The rigorous assessment of the storage stability of a drug product as it relates to its ability to provide a consistent delivery of cannabinoids is also necessary to provide expiration dates and storage conditions for the package labeling and/or product insert. In summary, all dosage formulations, including cannabis extracts, pills, lotions, injectables, and consumables, should be shown to be accurately formulated and consistently produced, and to have the delivery characteristics and stability attributes required for safety and efficacy over the lifetime of the product.

ADDITIONAL CONSIDERATIONS

It may not be intuitively obvious that an analytical method that has been validated for a cannabis-containing extract or "infused" beverage may not be applicable to a cannabinoid-containing cookie or gummy bear. However, as the complexity of the dosage formulation increases, it is increasingly likely that it will contain ingredients or excipients that can affect the recovery and accuracy of a particular analytical method developed and validated for a different dosage formulation. Therefore, analytical methods should be validated for each formulation and used to assure consistency in the constituents and their ability to serve as a reproducible dose delivery system. Why do not ethical pharmaceutical companies make their dosage formulations in several varieties of cookies, lollipops, and chocolate bars? Beyond the potential to be mistaken as foods, it could also be argued that the requirements for the control of the quality of the ingredients and excipients, and the determination of their compatibility and stability, would result in prohibitive production costs. Each unique product would not only require targeted methods that had been validated for constituents at their specified levels, but the final products would likely be tested for their disintegration or dissolution to ensure that the product released its active substances in a desirable and reproducible fashion. Thus, for FDA approval, the analytical methods would need to be developed, validated, and employed in formulation development, manufacturing, stability studies, dissolution testing, and nonclinical and clinical studies showing that they can be safe and effective drug products. In response, reputable pharmaceutical formulations are typically manufactured to contain the least number of ingredients and excipients required, in a limited number of doses and formulations, with validated analytical methods and well-characterized, consistent approaches for manufacture, packaging, and labeling. While this is not a requirement for all distributors of cannabis, either legal or illicit, it is a standard that ensures that the quality of a botanical drug substance or finished botanically derived drug product is established and maintained through standard conditions and expected duration of use.

Close-Up: Cannabis Quality Assurance?

Ryan Vandrey
Johns Hopkins University, Baltimore, MD

In 1996, California became the first state to pass legislation allowing residents to use cannabis for medical purposes. This was landmark legislation that has been followed by another 22 states and the District of Columbia. There have been growing pains and controversies, one of which was how to disentangle the conflict between federal and state law. Initially, the focal point in the media was on criminal cases where federal authorities arrested and/or confiscated property of individuals and businesses allegedly operating in compliance with the state law. However, perhaps a more pressing concern is the inability of federal regulators to oversee the large medical and nonmedical cannabis industries.

Living in Maryland, where the use of cannabis for any reason was and is illegal, it was difficult for me to directly observe the evolution of the "legal" market. So I would occasionally search Internet websites that describe cannabis products, offer menus for dispensaries in various locations, and comment on the expanding legal market. I was shocked by the variety of products on offer! I recall the online menu of a dispensary in California that offered 75 types of dried cannabis and over 150 other products, including oils, concentrates, edibles, creams, and transdermal patches. This was intriguing for a couple reasons. First, it was unclear why a dispensary would need to provide 75 strains. Were these really very different, or was it more of a marketing ploy? Second, I was curious whether what was being sold as a specific strain such as "Green Crack" or "OG Kush" in one dispensary was the same as that being sold under the same name elsewhere.

When travel took me to western states, I would often search out a dispensary and ask if I could come in and ask questions. Some were polite and welcoming, others showed me the door. On one visit to California, I spoke to a gentleman in the front office of a dispensary for about 30 minutes. Many of the their products were carefully labeled to indicate the amount/ concentration of THC and sometimes minor cannabinoids. I asked if they tested the products to ensure accuracy and he said no. He gave the same answer when I asked whether they did safety testing (eg, mold, pesticides). I asked how he knew the products they received from growers and manufacturers were reliable, safe, and accurately labeled. He replied that, if the product was not as advertised, the community would not stand for it and the company or individual would be forced out of business. I asked how many companies he worked with and whether he knew what they did for testing. He said the dispensary was a cooperative that worked with several people. When I asked how many cooperative members there were, he said he did not know. During our chat, I saw several people carrying duffel bags in and out. I left with the impression that this dispensary was nothing more than a consignment shop for anybody who wished to sell cannabis or cannabis-based products, and that there was no system of quality assurance whatsoever.

I immediately wanted to conduct a study to test the quality and label accuracy of products being sold in medical dispensaries. But ... there was a major hurdle. Purchasing, transporting, or otherwise possessing cannabis not obtained from the federal government was illegal. Though I receive cannabis for my research from the federal government, and though the DEA agents I spoke to about my proposal thought it was a great idea, there was no legal way to obtain cannabis products openly sold in states that allow medical use and have them evaluated in federally approved facilities. Frustrated, I tabled the idea.

That summer, at the annual meeting of the International Cannabinoid Research Society, I saw a presentation by Jeff Raber, who described his cannabis plant testing in California. His data showed significant variability in the chemotype (ie, concentrations of cannabinoids, terpenes, and flavonoids) of cannabis labeled as being of a certain strain. I explained my interest and we quickly agreed to partner on a project. Because he had already obtained and tested a large volume of plant material, we decided to focus on edibles. By using laboratories that he operated in Washington and California, we were able to do the study in accordance with state law and did not transport products across state lines. We selected edibles because they were growing in popularity, there was no natural ceiling on the amount of cannabinoids that could be put into a product, and manufacturing provides a lot of room for error. Thus, it was no surprise that we rarely found the measured quantities of THC and CBD that were listed on the label. What was most surprising was the variety of edibles available from so few dispensaries, and the range of doses. That there was a product with a listed content of 1000 mg of THC (it tested at 1236 mg) is simply astounding. To put that in perspective, the maximum recommended medical dose of THC (dronabinol) is 20 mg, and in my laboratory several daily cannabis users became sick or vomited after swallowing an 80 mg dose. Yes, this product was labeled as containing 20 doses, but it resembled a brownie and was no bigger than a typical single serving, there was no clear way of selecting one dose from it, and the THC was not evenly distributed.

As I write this, the State of Maryland is accepting applications to begin a medical cannabis program. On paper, the regulations for this program are very strict and appear to place special emphasis on product reliability and safety. However, it remains to be seen whether there is adequate expertise, standards for testing, and enforcement infrastructure. Without clear, consistent, enforced quality control standards, the current legal medical cannabis industry offers little consumer advantage over the black market, aside from the convenience and comfort of purchase at a retail store. Thus, there is a dire need for the development and implementation of standards for the cultivation, processing, labeling, and quality assurance testing of cannabis products. Federal agencies are much more capable of doing this, which highlights the importance of resolving the discrepancy between state and federal law on cannabis for medical or nonmedical reasons.

REFERENCES

1. Zuardi AW. History of cannabis as a medicine: a review. *Rev Bras Psiquiatr*. 2006;28:153−157.

2. Trofin IG, Dabija G, Vaireanu DI, Filipescu L. Long-term storage and cannabis oil stability. *Rev Chim*. 2012;63(3):293−297.

3. Gaoni Y, Mechoulam R. Hashish—VII. *Tetrahedron*. 1966;22(4):1481−1488.

4. Adams R, Pease DC, Cain CK, et al. Conversion of cannabidiol to a product with marihuana activity. A type reaction for synthesis of analogous substances. *J Am Chem Soc*. 1940;62 (8):2245−2246.

5. Watanabe K, Itokawa Y, Yamaori S, et al. Conversion of cannabidiol to Δ9-tetrahydrocannabinol and related cannabinoids in artificial gastric juice, and their pharmacological effects in mice. *Forensic Toxicol*. 2007;25(1):16−21.

6. Martin-Santos R, Crippa JA, Batalla A, et al. Acute effects of a single, oral dose of d9-tetrahydrocannabinol (THC) and cannabidiol (CBD) administration in healthy volunteers. *Curr Pharm Des*. 2012;18(32):4966−4979.

7. Manini AF, Yiannoulos G, Bergamaschi MM, et al. Safety and pharmacokinetics of oral cannabidiol when administered concomitantly with intravenous Fentanyl in humans. *J Addict Med*. 2015;9(3):204−210.

8. Vandrey R, Raber JC, Raber ME, Douglass B, Miller C, Bonn-Miller MO. Cannabinoid dose and label accuracy in edible medical cannabis products. *JAMA*. 2015;313 (24):2491−2493.

9. United States Food and Drug Administration. Warning letters and test results. <http://www.fda.gov/NewsEvents/PublicHealthFocus/ucm435591.htm>; 2015.

10. MacCoun RJ, Mello MM. Half-baked—the retail promotion of marijuana edibles. *N Engl J Med*. 2015;372(11):989−991.

11. Berger E. Legal marijuana and pediatric exposure pot edibles implicated in spike in child emergency department visits. *Ann Emerg Med*. 2014;64(4):A19−A21.

12. Potera C. Kids and marijuana edibles: a worrisome trend emerges. *Am J Nurs*. 2015;115 (9):15.

13. Organization WH. WHO guidelines on good agricultural and collection practices (GACP) for medicinal plants, <http://apps.who.int/iris/bitstream/10665/42783/1/9241546271.pdf>; 2003 Accessed 16.11.15.

14. Agency EM. Guideline on good agricultural and collection practice for starting materials of herbal origin. <http://www.ema.europa.eu/docs/en_GB/document_library/Scientific_guideline/2009/09/WC500003362.pdf>; 2006 Accessed 16.11.15.

15. Potter DJ. Cannabis horticulture. In: Pertwee RG, ed. *Handbook of Cannabis*. New York, NY: Oxford University Press; 2014:65−88.

16. Stott CG, Guy GW. Cannabinoids for the pharmaceutical industry. *Euphytica*. 2004;140 (1−2):83−93.

17. Potter DJ. A review of the cultivation and processing of cannabis (*Cannabis sativa* L.) for production of prescription medicines in the UK. *Drug Test Anal*. 2014;6(1−2):31−38.

18. Clarke RC. *Marijuana Botany: An Advanced Study: The Propagation and Breeding of Distinctive Cannabis*. Berkeley, CA: Ronin Publishing; 1981.

19. Upton R, Craker L, ElSohly M, Romm A, Russo E, Sexton M. *Cannabis Infloresence*. Scotts Valley, CA: American Herbal Pharmacopoeia; 2013.

20. Taschwer M, Schmid MG. Determination of the relative percentage distribution of THCA and Δ9-THC in herbal cannabis seized in Austria—Impact of different storage temperatures on stability. Forensic Sci Int. 2015;254:167−171.

21. Organization WH.. WHO guidelines on good manufacturing practices (GMP) for herbal medicines. <http://apps.who.int/medicinedocs/documents/s14215e/s14215e.pdf>; 2007 Accessed 16.11.15.

22. Administration USFaD. In: Research CfDEa, ed. *Guidance for Industry: Botanical Drug Development (Draft Guidance)*. Silver Spring, MD: Office of Communications, Division of Drug Information; 2015.

23. Gieringer D, St Laurent J, Goodrich S. Cannabis vaporizer combines efficient delivery of THC with effective suppression of pyrolytic compounds. *J Cannabis Ther*. 2004;4(1):7−27.

24. Pomahacova B, Van Der Kooy F, Verpoorte R. Cannabis smoke condensate III: the cannabinoid content of vaporised *Cannabis sativa* cannabinoid content of vaporised *Cannabis sativa*. *Inhal Toxicol*. 2009;21(13):1108−1112.

25. Hartman RL, Brown TL, Milavetz G, et al. Controlled cannabis vaporizer administration: blood and plasma cannabinoids with and without alcohol. *Clin Chem*. 2015;61(6):850−869.

26. Budney AJ, Sargent JD, Lee DC. Vaping cannabis (marijuana): parallel concerns to e-cigs? *Addiction*. 2015;110:1699–1704.

27. Giroud C, de Cesare M, Berthet A, Varlet V, Concha-Lozano N, Favrat B. E-Cigarettes: a review of new trends in cannabis use. *Int J Environ Res Public Health*. 2015;12 (8):9988–10008.

28. Agu RU, Ugwoke MI. In vitro and in vivo testing methods for respiratory drug delivery. *Expert Opin Drug Deliv*. 2011;8(1):57–69.

29. Darquenne C. Aerosol deposition in health and disease. *J Aerosol Med Pulm Drug Deliv*. 2012;25(3):140–147.

30. Pilcer G, Amighi K. Formulation strategy and use of excipients in pulmonary drug delivery. *Int J Pharm*. 2010;392(1–2):1–19.

31. Bonuccelli CM. Stable solutions for marijuana analysis. *J Pharm Sci*. 1979;68(2):262–263.

32. Smith RN, Vaughan CG. The decomposition of acidic and neutral cannabinoids in organic solvents. *J Pharm Pharmacol*. 1977;29(5):286–290.

33. Turner CE, Henry JT. Constituents of *Cannabis sativa* L. IX: stability of synthetic and naturally occurring cannabinoids in chloroform. *J Pharm Sci*. 1975;64(2):357–359.

34. Flora KP, Cradock JC, Davignon JP. Determination of delta 9-tetrahydrocannabinol in pharmaceutical vehicles by high-performance liquid chromatography. *J Chromatogr*. 1981;206(1):117–123.

35. Fairbairn JW, Liebmann JA, Rowan MG. The stability of cannabis and its preparations on storage. *J Pharm Pharmacol*. 1976;28(1):1–7.

36. Lewis GS, Turner CE. Constituents of *Cannabis sativa* L. XIII: stability of dosage form prepared by impregnating synthetic (−)-delta 9-trans-tetrahydrocannabinol on placebo cannabis plant material. *J Pharm Sci*. 1978;67(6):876–878.

37. Narayanaswami K, Golani HC, Bami HL, Dau RD. Stability of *Cannabis sativa* L. samples and their extracts, on prolonged storage in Delhi. *Bull Narc*. 1978;30(4):57–69.

38. Taschwer M, Schmid MG. Determination of the relative percentage distribution of THCA and Delta(9)-THC in herbal cannabis seized in Austria—impact of different storage temperatures on stability. *Forensic Sci Int*. 2015;254:167–171.

39. Lindholst C. Long term stability of cannabis resin and cannabis extracts. *Aust J Forensic Sci*. 2010;42(3):181–190.

CHAPTER 6

The Roles of Research and Regulation

FROM HERBAL MEDICINES TO CONTROLLED SUBSTANCES

Natural products have been used throughout history as food substances, incense, herbs, and spices, and as intoxicants and herbal medicines.[1] The cultivation and use of cannabis or hemp for one or more of these purposes appears to date back at least 10,000 years to the very emergence of agriculture. In western cultures the popularity of medicinal cannabis and cannabis-derived tinctures peaked in the late nineteenth century. Then came a change in perspective and a preference for well-defined drug products as opposed to herbal remedies, giving rise to modern pharmaceutical science and, concurrently, national and international drug control measures. Interest in the evolving technology and science around medicinal chemistry and pharmacology, and the isolation and testing of the active constituents from natural products, diminished interest in and use of complex, unrefined herbal medicines.[2]

As a result of increasing concern over substance abuse, the 1961 UN Single Convention on Narcotic Drugs subjected cannabis and cannabis derivative products to special measures of control, and parties could ban their use altogether.[3] Psychoactive substances were categorized according to the perceived risks associated with their abuse liability and benefits based on their therapeutic utility. The placement of the cannabis plant and its resin in both Schedules I and IV, and extracts and tinctures in Schedule I, essentially prohibited their legal use. In 1971 the UN Convention on Psychotropic Substances proposed a new system of classification to address synthetic drugs, and Δ^9-THC and its isomers were again classified in the most restricted Schedule (I). Cannabidiol (CBD) was not controlled under the Psychotropic Convention, but the United States and other nations elected to control it and other pure cannabinoid preparations. For example, the US Controlled Substances Act of 1970 placed "all parts of the plant *Cannabis sativa* L., whether growing or not, the seeds thereof; the resin extracted from any part of such plant; and every compound,

The Analytical Chemistry of Cannabis. DOI: http://dx.doi.org/10.1016/B978-0-12-804646-3.00006-0

manufacture, salt, derivative, mixture or preparation of such plant, seeds or resin" in Schedule I. With the exception of Marinol (dronabinol, synthetic $(-)$-*trans*-Δ^9-THC), which is Schedule III, all non-THC pure cannabinoids, including CBD, are Schedule I in the United States. With its current scheduling cannabis and cannabis-derived dosage formulations can only be legally obtained from the National Institute on Drug Abuse (NIDA); procurement from other sources, including medicinal or recreational use dispensaries and programs under select state regulations, is still in violation of US and international treaty laws.

IMPLICATIONS TO CONSUMERS

The existing patchwork nature of international, federal, state, and regional regulations and enforcement over cannabis and cannabis-derived drug substances (Fig. 6.1) presents an array of challenges for patients, recreational users, entrepreneurs, drug manufacturers, researchers, customs agents, law enforcement officers, and regulatory agencies. The US, the federal government considers cannabis a Schedule I controlled substance and Δ^9-THC in the form of Marinol a Schedule III controlled substance, while certain states and districts have passed legislation allowing medicinal or recreational open access and use of these or similar substances to adults. As a result, trafficking

Legend:
- Legal
- Decriminalized
- Legal for medicinal use
- Unenforced
- Illegal
- No information

Figure 6.1 Global variability in the legality of cannabis use in 2015. Note that the legal status of cannabis is evolving rapidly, and in some areas the status is unknown or sufficiently unclear that "no information" was used. It is also important to note that within the United States, cannabis and its constituents (including pure Δ^9-THC and CBD) are still Schedule I controlled substances under the US Controlled Substances Act (1970).

of substances from noncontrolled to controlled regions is widespread. Some states that consider cannabis use illegal have resorted to litigation against other states providing legal access for medical or recreational use because of the increase in transportation of controlled substances across their state lines. Patients may need to move or cross state lines to gain access to medicinal products. The long half-life of Δ^9-THC and other phytocannabinoids further confounds the issues associated with recreational intoxication or medicinal use and law enforcement in the existing global framework. In principal, a person could drive to one jurisdiction and use cannabis legally, and then several days later return to an area of strict control and enforcement, test positive for use, and suffer considerable legal consequences. This situation is exacerbated by the increasing availability of synthetic cannabinoid "designer drugs" such as JWH-018 and other cannabimimetics, which enable people to become intoxicated and avoid detection, but represent exceptional risks to public health because of their structural novelty and lack of quality control and pharmacological and toxicological information on safety and efficacy during acute or chronic use.

IMPLICATIONS TO SUPPLIERS

The current legal framework also challenges individual entrepreneurs and large-scale pharmaceutical manufacturers who wish to address the public's desire or need for access to improved recreational or medicinal cannabis or phytocannabinoid-containing products. In the United States, for instance, entrepreneurs in states where cannabis is legal for use must currently deal in cash and face difficulty establishing bank accounts or insurance to protect their earnings and investments. In addition, psychoactive cannabis or cannabinoid-containing products that are regulated (eg, Schedule I or II substances such as Δ^9-THC- and CBD-containing products) are likely to be of limited interest to large-scale pharmaceutical companies because of the barriers to licensing and sales and the significant costs of compliance with existing federal or international controls. In 2014 the US National Institute on Drug Abuse (NIDA) increased the annual production of cannabis at the University of Mississippi from 18 to 600 kg, as it is the only source considered legal by federal authorities (described in detail below).

IMPLICATIONS TO RESEARCHERS

US researchers find themselves in a particularly problematic position. Many can only watch from the sidelines as real-world experimentation with cannabis and cannabinoids is conducted without proper controls, reliable or well-characterized and consistent dose formulations, appropriately informed consent, and other procedural attributes required to adequately assess treatments and outcomes for safety and efficacy. For example, if an investigator wanted to acquire drug substances from dispensaries in Colorado and submit them for chemical analysis in North Carolina, shipment to North Carolina would be a violation of US law. For the materials to be packaged and shipped in accordance with DEA regulations, the intended recipient would need to request that a DEA-222 form be sent from the dispensary. But this is not likely to happen as long as the DEA considers the only legal supply of marijuana in the United States to be that grown at the University of Mississippi and made available through NIDA. Thus, researchers are limited in what can be done using material that is available through state-approved dispensaries and distribution programs. And because Congress voted to prohibit the federal government from raiding such facilities, the products are not available for NIDA or its researchers to study.

To facilitate cannabis investigations, NIDA has a long history of providing consistent cannabis-derived dosage formulations to patients through the Compassionate Investigational New Drug Program, and to researchers. Bulk plant material, cannabis cigarettes, cannabis extracts, and other dosage formulations in a variety of phytocannabinoid strengths (including ethanol extracted "placebo" marijuana) can be requested through the NIDA Drug Supply Program (DSP) for use in preclinical and clinical studies. NIDA operates its DSP in accordance with DEA regulations for controlled substances. Its cannabis supply is grown, harvested, processed, analyzed for Δ^9-THC and other cannabinoids, and stored under controlled conditions to preserve purity and stability at the University of Mississippi. Batches of plant material and extracts are sent with a certificate of analysis to RTI International in Research Triangle Park, North Carolina, where they are quarantined, further analyzed, and stored for release and distribution. Alternatively, the material can be analytically tested and released for blending or dosage formulation and manufacturing. NIDA's finished cannabis and cannabis-based dosage formulations are fully characterized and released for distribution, stored in secure DEA registered facilities under controlled conditions, and distributed only at the authorization of NIDA.

Persons in the United States who have funding with the National Institutes of Health (NIH) and wish to conduct research using NIDA's drug substances and formulations must first obtain a special DEA registration under the Controlled Substances Act, and then submit a request for consideration to NIDA along with other required documentation about their intended use. To facilitate the use of these materials, NIDA has compiled a Drug Master File with the Food and Drug Administration (FDA) that provides the FDA with information on the seed selection, growth, harvesting, manufacturing, analytical characterization, and stability testing for each batch of its cannabis and cannabis-derived dosage formulations. Investigators who desire access to human use material from the NIDA DSP must have an approved Investigational New Drug Application (IND) for the conduct of clinical studies that references the NIDA Drug Master File. If the material is to be used for other than clinical human research, the request is simply forwarded to NIDA's Office of the Director for review and recommendation. Researchers without an NIH grant are subject to additional scientific review. For non-grantees using cannabis and cannabis dosage formulations from the NIDA DSP in human research, an additional review is required by the US Department of Health and Human Services, Public Health Services. Foreign applicants must provide necessary documentation that controlled substances, research chemicals, or marijuana cigarettes being requested are permitted for import to their countries. When NIDA approves and authorizes the distribution of a Schedule I controlled substance to a researcher or patient, the recipient must have the appropriate DEA registration and associated drug code, and submit a DEA-222 official order form. When appropriate documentation and approval has been received, cannabis and cannabinoid-containing dosage formulations are released for use in preclinical and clinical studies or for medicinal use under the Compassionate Use Act. Each batch of cannabis or cannabis-derived dosage formulation has a certificate of analysis or data sheet that is provided to the researcher or end-user containing information on the particular batch along with recommendations for its proper handling, storage, and use.

Despite the provision to researchers of access to consistent and analytically well-characterized cannabis and cannabis-derived dosage formulations, the documentation and approvals needed to receive cannabis from the US government are often described as time-consuming and difficult. Sara Reardon in *Nature News*: "Residents of 23 US states can buy medical marijuana to treat everything from cancer pain to anxiety, but US scientists must wade through onerous paperwork to score the drug for study."[4] The range of available cannabis varieties and dosage formulations available from NIDA is limited. For example, researchers and individuals receiving medicinal cannabis through the Compassionate Investigational New Drug program have historically been limited to placebo and Δ^9-THC-containing cannabis varieties, and the potency of this material was often limited to Δ^9-THC levels below those available in the illicit market, the intent being to avoid or limit tolerance or dependence. While this was often portrayed negatively, it was also recognized that the interest of the NIDA DSP was likely to be limited to the provision of well-characterized and consistent dosage formulations to be used with appropriate controls and procedures to ensure investigations were conducted in a safe and reproducible fashion. In 2014 NIDA increased its spending on research marijuana by 50%, and the crop harvested in late 2015 year included two new strains. One has low concentrations of tetrahydrocannabinols (THCs), marijuana's primary active ingredients, but high levels of CBD; the second has relatively balanced levels of the two. Dr Nora Volkow, the Director of NIDA, was quoted as saying "We want to be able to evaluate the claims that marijuana is therapeutically beneficial" and to explore treatments for addiction.[4] NIDA's DSP also supplies reference materials and hard to procure standards of cannabinoids and their metabolites, as well as a selection of radioisotope- or stable isotope–labeled cannabinoid analogs for use in research, and provides a contract mechanism with RTI International for the synthesis of compounds as requested by researchers. The DEA still requires researchers and the NIDA synthesis and drug supply programs to adhere to all of the typical labeling, inventory, distribution controls, and the meticulous record-keeping required for other chemicals falling under this schedule; whereas products distributed through state programs are not required or penalized for failing to follow these requirements. Moreover, individual cannabinoid researchers at US universities are often required to apply for and obtain their own DEA registrations and provide for the secure storage and inventory management systems necessary for Schedule I controlled substances.

IMPLICATIONS TO REGULATORS

Resources continue to be consumed by international agencies such as the United Nationals Drug Control and the US FDA and DEA as they determine how to address (or in some instances ignore) the current situation to provide for and protect the public's health and well-being. For example, the growing unwillingness among an increasing number of nations to apply the strictly prohibitionist interpretation of the UN drug control conventions with respect to medicinal cannabis suggests that some form of treaty revision and legislative reform of national drug policies should be considered.[5] By revising drug policies, the FDA and other regulatory agencies might be further empowered to oversee, improve, and ensure the quality of cannabis and cannabinoid-containing products for use in medicinal, recreational, or religious practices. The development and use of herbal formulations of novel synthetic cannabinoids and other new illicit psychoactive substances of abuse is another area of public concern requiring unique approaches to legal frameworks and regulatory actions (see Glass Close-Up: The New Zealand Experience).

In addition to the regulation of the naturally occurring constituents in cannabis and synthetics, the use of pesticides and growth regulators during medicinal cannabis cultivation can also be a considerable challenge for regulatory agencies. No specific pesticides have been registered for use with cannabis in the United States, and this seems unlikely to occur soon given the illegal status of cannabis under federal law. The absence of approved pesticides for use in the United States creates a dilemma for the legal medical and recreational cannabis industry that seeks to balance lawful compliance with pesticide use, worker health, and public health, particularly since the absence of approved products for use on cannabis may result in consumer exposures to otherwise more hazardous pesticides or higher residue levels than would occur if regulatory guidance and specifications existed. Indeed, despite the lack of product registrations for cannabis, illegal pesticide use has been verified through residue monitoring, enforcement activities, and legislative testimony.[6]

Close-Up: The New Zealand Experience: A Novel Approach to Regulating Synthetic Cannabinoids

Michelle Glass
Department of Pharmacology, University of Auckland, Auckland, New Zealand

Similar to most developed countries, New Zealand has seen a dramatic increase in the number and availability of "legal highs" over the past decade. These compounds, initially marketed as natural and herbal, were soon realized by the international community to contain highly potent synthetic cannabinoids.[1]

As community concerns about new psychoactive products grew, the New Zealand Government, along with many international regulators, placed a series of temporary bans on products. These bans were based on the identified chemical structure of the psychoactive ingredient of each product. Between 2005 and 2013, 35 distinct chemical structures were identified and banned. However, such an approach was doomed to failure, with banned compounds being rapidly replaced by alternative, structurally distinct compounds, often with no change to packaging or product names.

The New Zealand Government thus looked for a different approach and passed a new law in 2013. This legislation was novel in that, rather than banning all substances, it acknowledged the demand for such products, and innovatively shifted the onus to manufacturers to prove that the compounds were safe. This "Psychoactive Substances Act" (PSA) has as its main purpose, "[to] regulate the availability of psychoactive substances in New Zealand to protect the health of, and minimize harm to, individuals who use psychoactive substances."[2]

The PSA therefore requires producers to establish through preclinical and clinical trials, to the satisfaction of an expert advisory panel, that their products "pose no more than a low risk of harm to individuals who use it." Nowhere does the PSA describe what a "low risk of harm" would look like, but the Ministry of Health was tasked with outlining this in subsequent guidelines. The PSA also put significant restrictions around advertising of psychoactive substances, and criminalized the sale and supply, as well as purchase, of PSA-approved substances for individuals aged below 18 years. Furthermore, in an attempt to limit the exposure of youth, it specified that products could not be sold in convenience stores, petrol stations, supermarkets, or liquor outlets.

The bill provides a broad definition of a psychoactive substance: "a substance, mixture, preparation, article, device, or thing that is capable of inducing a psychoactive effect (by any means) in an individual who uses the psychoactive substance".[2] Thus it casts a wide net not only on synthetic cannabinoids but also on any other classes of compounds that could emerge in coming years. The Act specifically excludes alcohol, tobacco, food, medicines, and, importantly, drugs currently scheduled under the Misuse of Drugs Act. As marijuana is scheduled under Class C of the Act, and more potent forms, such as hashish and cannabis oil, are scheduled under Class B,[3] the act therefore specifically excluded any attempt to use the legislation to legalize cannabis.

In the initial stages of the PSA's introduction, products that were already on the market could legally remain so under an "interim regime" while the testing regime was developed. Substances identified as causing harm during this period, usually on the basis of reports to the New Zealand Poisons Centre or hospital admissions, were withdrawn from the market by the Ministry of Health. Five products were removed in late 2013 and six in April 2014. However, the products were proving popular, and a large number of media reports profiled wide ranging adverse effects from psychotic episodes, seizures, vomiting, and panic attacks. A public backlash against the legal sale of synthetic cannabinoids developed, and in April 2014 the government announced a decision to end the interim phase of the PSA. This decision removed the remaining 36 products on sale, effective as of May 8, 2014.

Thus currently there are no approved psychoactive products legally available for sale in New Zealand, and no products will be available until they have been fully approved under the Regulations. The regulatory environment was further complicated by the decision, again in response to public pressure, to ban the use of animals in research geared toward establishing the safety of a psychoactive substance (perhaps notably, 2014 was an election year in New Zealand). Thus while the guidelines now lay out a process for establishing "no more than a low risk of harm" which progresses from preclinical to clinical testing, they acknowledge that, "at this time, the OPSRA is unaware of any suitable nonanimal alternatives for assessing the pharmacokinetics, metabolism, reproductive toxicity, or addiction potential of a substance. These are key requirements in assessing risk." Thus it is near impossible that a substance could currently work its way through the Act's requirements to be made legal. As might have been expected, the media are reporting an increase in black market synthetic cannabinoid sales,[4] and while there has been some suggestion that the number of patients being admitted with synthetic cannabinoid induced psychotic episodes had decreased,[5] there have continued to be reports of hospital admissions due to adverse responses to synthetic cannabinoids.[6]

So this bold and somewhat ground-breaking legislation appears to have been rendered impotent, and New Zealand is almost back to where it was before legislation was introduced, with a range of unregulated illegal products. The legislation contains a requirement for it to be reviewed by the Parliament by 2018, and it will be interesting to see what has been learnt from this process.

http://www.fr-online.de/rhein-main/frankfurt-gefaehrlicher-kick-mit--spice-,1472796,3375090.html;   http://www.sciencemediacentre.co.nz/2011/07/11/what-is-in-legal-highs-new-data-from-esr/

http://www.legislation.govt.nz/act/public/2013/0053/latest/whole.html

http://www.legislation.govt.nz/act/public/1975/0116/latest/DLM436101.html

http://www.nzherald.co.nz/nz/news/article.cfm?c_id=1&objectid=11488836;   http://www.nzherald.co.nz/bay-of-plenty-times/news/article.cfm?c_id=1503343&objectid=11485539

http://www.nzherald.co.nz/nz/news/article.cfm?c_id=1&objectid=11448901

http://www.nzherald.co.nz/nz/news/article.cfm?c_id = 1&objectid=11454095;   http://www.nzherald.co.nz/bay-of-plenty-times/news/article.cfm?c_id=1503343&objectid=11484248

REFERENCES

1. Booth M. *Cannabis: A History*. New York, NY: St. Martin's Press; 2003.

2. Mead AP. International control of cannabis. In: Pertwee RG, ed. *Handbook of Cannabis*. New York, NY: Oxford University Press; 2014:44–64.

3. Stott CG, Guy GW. Cannabinoids for the pharmaceutical industry. *Euphytica*. 2004;140 (1–2):83–93.

4. Reardon S. Marijuana gears up for production high in US labs. *Nature*. 2015;519 (7543):269–270.

5. Bewley-Taylor DR. Towards revision of the UN drug control conventions: harnessing like-mindedness. *Int J Drug Policy*. 2013;24(1):60–68.

6. Stone D. Cannabis, pesticides and conflicting laws: the dilemma for legalized States and implications for public health. *Regul Toxicol Pharmacol: RTP*. 2014;69(3):284–288.

CHAPTER 7

The Future of Cannabinoid Therapeutics

The pharmacologic and therapeutic effects of drugs are dependent on formulation and dose, route of administration, and patient characteristics. People now have access to more varieties of cannabis and cannabis-derived formulations for a wider range of therapeutic indications or recreational use than ever before. For example, dronabinol (Marinol) and nabilone (Cesamet) are orally administered, prescription-based, Δ^9-THC-containing pharmaceutical preparations approved by the FDA and other regulatory agencies for the treatment of nausea and vomiting associated with cancer chemotherapy and of anorexia and cachexia in patients with acquired immune deficiency syndrome and other disorders. Chronic pain[1] and muscle spasm in multiple sclerosis[2] are also recognized as indications that can be treated effectively with these and other cannabis-based therapeutics such as nabiximols (Sativex). Nabiximols has not been approved in the United States but is available in other countries as a prescription sublingual spray containing a cannabis extract with Δ^9-THC and the nonpsychoactive compound cannabidiol (CBD) as its primary active ingredients. CBD and high-CBD cannabis strains are also of clinical interest for their purported antiseizure activity and other pharmacological effects. However, having access to a variety of drug substances for a wide range of conditions is not a guarantee of their safety or efficacy, and may place people at considerable health risk if improperly formulated or used.

The accumulating experimental evidence and public acceptance of the therapeutic utility of cannabis and cannabis-derived drug products appears to be particularly vulnerable to exploitation and misuse in the current political environment, where regulation and control hamper research and pharmaceutical development, but laws remain widely unenforced or ignored in the nonfederally regulated setting. The list of medical indications that are being claimed for cannabis and cannabinoids is expanding at a rate that far exceeds scientific review or

The Analytical Chemistry of Cannabis. DOI: http://dx.doi.org/10.1016/B978-0-12-804646-3.00007-2

evaluation as fact or fiction. For instance, based on anecdotal reports, limited experimental evidence obtained in open-label trials, and widespread media coverage, CBD is promoted as a treatment for pediatric epilepsy, especially Dravet syndrome. However, a recent study that supports a potential role for CBD in the treatment of refractory childhood epilepsy including intractable seizures and Lennox–Gastaut syndrome emphasized that appropriately controlled clinical trials are still required to adequately establish efficacy and safety.[3] Similarly, an increasing number of states have approved posttraumatic stress disorder (PTSD) as a qualifying condition for medical marijuana, although evidence of efficacy is limited. It appears that in some instances marijuana may actually worsen PTSD symptoms or nullify the benefits of specialized intensive treatments, such that cessation or prevention of use in some patients may be warranted.[4]

There is a great need for research in specific areas of cannabis and cannabinoid therapeutics, such as the definition of chemical constituents and treatment modalities and improved methods for assessing efficacy and outcomes. It has been only a little over 50 years since Roger Adams[5], Raphael Mechoulam,[6] and others opened the world's eyes to the chemistry of cannabis and the cannabis-like effects associated with tetrahydrocannabinols. The structures of CB1 and CB2 receptors and endocannabinoids were only discovered[7,8] and cloned[9,10] in the 1980s and 1990s. Dronabinol, nabilone, and nabiximols are still industry standards of efficacy and safety for their specific indications. And federal regulations are becoming less prohibitionist and more accepting (or forgiving) of the use of cannabis products as medicine or intoxicant. However, there are limits to access even in free markets, and there are societal expectations for regulation to be placed over pharmaceutical substances that will ensure that the public health is well served and protected. It is unethical to promote cannabinoids for indications that have not been fully evaluated and documented, or to distribute drug substances in untested drug delivery devices without clear labeling of ingredients, secure packaging, and specific consumer information and instructions. E-cigarette drug delivery devices may be less harmful than smoking cannabis, but the inhalation of concentrated formulations of cannabis may have abuse and dependence liability above that of smoking the herb and could facilitate initiation or continued use under situations where it would be inappropriate but undetected. The United States has seen an increased rate of admissions to substance treatment facilities for cannabis use disorders, where it is the most commonly identified illicit substance.[11,12] In addition to the potential risks associated with smoking or e-cigarette initiation, the epidemiological literature in the past two decades indicates that cannabis use increases the risk of motor-vehicle accidents, and there is a consistently found association between regular cannabis use in youth and poor psychosocial outcomes and mental health in adulthood.[13] Therefore, NIDA and the DEA should remain involved in the research and regulatory framework going forward to further our understanding of abuse liability and other potential hazards, to limit diversion, and to control public access to structurally unique synthetic cannabinoids (eg, JWH-018) and novel herbal formulations until sufficient information is obtained to justify their use and distribution in human and animal medicine.

The FDA's recent issuance of Draft Guidance for Industry on Botanical Drug Dependence is likely to be the first step among many taken to ensure that the public only has access to cannabis-based therapeutics that have proven to be safe and effective. The American Herbal Pharmacopoeia, the American Herbal Products Association, the Natural Products Association, and other organizations and researchers involved with natural products or dietary supplements are also contributing to the knowledge base and best practices. Dietary supplement manufacturers must register their facilities with the FDA, and manufacturers and distributors must make sure that all claims and information on the product label and in other labeling are truthful and not misleading. Under FDA regulations at 21 CFR part 111, all domestic and foreign companies that manufacture, package, label, or hold dietary supplements, including parties involved with testing, quality control, and distribution in the United States, must comply with the Dietary Supplement Current Good Manufacturing Practices (CGMPs) for quality control. In addition, the manufacturer, packer, or distributor whose name appears on the label of a dietary supplement is required to submit to the FDA all serious adverse event reports associated with use. The FDA recently assumed regulatory authority over the tobacco industry. This may enable comparison and extrapolation to be made to cannabis products and markets. In this model, entrepreneurs, pharmaceutical companies, researchers, and regulators help to define best practices and establish areas where research is needed, further product development or refinement and standardization are warranted, product liabilities need to be eliminated, and funding prioritized. It seems likely that the FDA will assume a more prevalent role, and undertake a similar approach, in furthering the research and regulation of cannabis-derived therapeutics.

In summary, the results of scientific research must guide the decisions for laws and medical use of marijuana in the future. Analytical chemistry will continue to play a critical role in this process, particularly the characterization of cannabis and its formulations, and the study of physiological processes involved in health and disease states as they are affected by cannabinoid-containing therapeutics. For example, metabolomics, proteomics, and genomic approaches can be applied to both plant and patient to facilitate a broader understanding of the biochemical basis of experimental therapeutics derived from cannabis. More importantly, analytical chemistry must be applied to products that are distributed for research, recreation, or medication to substantiate their quality and establish the relationship between chemical exposure profiles and health outcomes. In this fashion, knowledge can be turned into practice to better serve the human condition.

REFERENCES

1. Hill KP. Medical marijuana for treatment of chronic pain and other medical and psychiatric problems: a clinical review. *JAMA*. 2015;313(24):2474–2483.

2. Sastre-Garriga J, Vila C, Clissold S, Montalban X. THC and CBD oromucosal spray (Sativex (R)) in the management of spasticity associated with multiple sclerosis. *Expert Rev Neurother*. 2011;11(5):627–637.

3. Hussain SA, Zhou R, Jacobson C, et al. Perceived efficacy of cannabidiol-enriched cannabis extracts for treatment of pediatric epilepsy: a potential role for infantile spasms and Lennox-Gastaut syndrome. *Epilepsy Behav*. 2015;47:138–141.

4. Wilkinson ST, Stefanovics E, Rosenheck RA. Marijuana use is associated with worse outcomes in symptom severity and violent behavior in patients with posttraumatic stress disorder. *J Clin Psychiatry*. 2015;76(9):1174–1180.

5. Adams R, Loewe S, Pease DC, et al. Structure of cannabidiol. VIII. Position of the double bonds in cannabidiol. Marihuana activity of tetrahydro-cannabinols [1]. *J Am Chem Soc*. 1940;62(9):2566–2567.

6. Gaoni Y, Mechoulam R. Hashish—VII. *Tetrahedron*. 1966;22(4):1481–1488.

7. Devane WA, Dysarz Iii FA, Johnson MR, Melvin LS, Howlett AC. Determination and characterization of a cannabinoid receptor in rat brain. *Mol Pharmacol*. 1988;34(5):605–613.

8. Devane WA, Hanus L, Breuer A, et al. Isolation and structure of a brain constituent that binds to the cannabinoid receptor. *Science*. 1992;258(5090):1946–1949.

9. Matsuda LA, Lolait SJ, Brownstein MJ, Young AC, Bonner TI. Structure of a cannabinoid receptor and functional expression of the cloned cDNA. *Nature*. 1990;346(6284):561–564.

10. Munro S, Thomas KL, Abu-Shaar M. Molecular characterization of a peripheral receptor for cannabinoids. *Nature*. 1993;365(6441):61–65.

11. Borgelt LM, Franson KL, Nussbaum AM, Wang GS. The pharmacologic and clinical effects of medical cannabis. *Pharmacother: J Hum Pharmacol Drug Ther*. 2013;33(2):195–209.

12. Danovitch I, Gorelick DA. State of the art treatments for cannabis dependence. *Psychiatr Clin North Am*. 2012;35(2):309–326.

13. Hall W. What has research over the past two decades revealed about the adverse health effects of recreational cannabis use? *Addiction*. 2015;110(1):19–35.

Printed in the United States
By Bookmasters